SpringerBriefs in Applied Sciences and Technology

SpringerBriefs present concise summaries of cutting-edge research and practical applications across a wide spectrum of fields. Featuring compact volumes of 50 to 125 pages, the series covers a range of content from professional to academic.

Typical publications can be:

- A timely report of state-of-the art methods
- An introduction to or a manual for the application of mathematical or computer techniques
- A bridge between new research results, as published in journal articles
- A snapshot of a hot or emerging topic
- An in-depth case study
- A presentation of core concepts that students must understand in order to make independent contributions

SpringerBriefs are characterized by fast, global electronic dissemination, standard publishing contracts, standardized manuscript preparation and formatting guidelines, and expedited production schedules.

On the one hand, **SpringerBriefs in Applied Sciences and Technology** are devoted to the publication of fundamentals and applications within the different classical engineering disciplines as well as in interdisciplinary fields that recently emerged between these areas. On the other hand, as the boundary separating fundamental research and applied technology is more and more dissolving, this series is particularly open to trans-disciplinary topics between fundamental science and engineering.

Indexed by EI-Compendex, SCOPUS and Springerlink.

Azlina Idris CEng · Aidatul Julia Abd Jabar ·
Wan Norsyafizan W. Muhamad

MIMO-OFDM Systems with Diversity Technique

PAPR Reduction

Azlina Idris CEng
School of Electrical Engineering
Universiti Teknologi Mara
Shah Alam, Selangor, Malaysia

Aidatul Julia Abd Jabar
School of Electrical Engineering
Universiti Teknologi Mara
Shah Alam, Selangor, Malaysia

Wan Norsyafizan W. Muhamad
School of Electrical Engineering
Universiti Teknologi Mara
Shah Alam, Selangor, Malaysia

ISSN 2191-530X   ISSN 2191-5318   (electronic)
SpringerBriefs in Applied Sciences and Technology
ISBN 978-981-96-1000-6   ISBN 978-981-96-1001-3   (eBook)
https://doi.org/10.1007/978-981-96-1001-3

© The Editor(s) (if applicable) and The Author(s), under exclusive license to Springer Nature Singapore Pte Ltd. 2025

This work is subject to copyright. All rights are solely and exclusively licensed by the Publisher, whether the whole or part of the material is concerned, specifically the rights of translation, reprinting, reuse of illustrations, recitation, broadcasting, reproduction on microfilms or in any other physical way, and transmission or information storage and retrieval, electronic adaptation, computer software, or by similar or dissimilar methodology now known or hereafter developed.
The use of general descriptive names, registered names, trademarks, service marks, etc. in this publication does not imply, even in the absence of a specific statement, that such names are exempt from the relevant protective laws and regulations and therefore free for general use.
The publisher, the authors and the editors are safe to assume that the advice and information in this book are believed to be true and accurate at the date of publication. Neither the publisher nor the authors or the editors give a warranty, expressed or implied, with respect to the material contained herein or for any errors or omissions that may have been made. The publisher remains neutral with regard to jurisdictional claims in published maps and institutional affiliations.

This Springer imprint is published by the registered company Springer Nature Singapore Pte Ltd.
The registered company address is: 152 Beach Road, #21-01/04 Gateway East, Singapore 189721, Singapore

If disposing of this product, please recycle the paper.

Preface

The world has witnessed the rapid evolution of wireless communication systems as the demand for information and entertainment is growing extremely at a rapid pace regardless of the users' location. This trend continues to accelerate today and in the near future. Users embrace this new technological advancement and demand an efficient, reliable, and high-speed wireless communication network to satisfy their needs. One of the prominent solutions that could meet this demand is Orthogonal Frequency Division Multiplexing (OFDM) and Multiple-Input Multiple-Output (MIMO) technology. In addition, MIMO-OFDM is combined to promote the benefits of high-performance systems and exploitation of the multipath diversity, increasing data rates and link reliability. Despite all the advantages offered by the OFDM, it still suffers from a major problem which is the high peak-to-average power ratio (PAPR), and this problem continues to exist in MIMO-OFDM systems. The high PAPR in OFDM causes the efficiency of the high-power amplifier (HPA) to decrease and increase power consumption. The high PAPR becomes a serious issue and the main barrier to the adaptation of OFDM multicarrier transmission in some wireless systems.

I would like to express my gratitude to Prof. Ir. Dr. Hamidah Mohd Saman for their insightful feedback and unwavering support throughout the writing and editing of this book. Special thanks to Associate Professor Dr. Rohana Hassan and Dr. Arni Munira Markom for their expertise and dedication, which were essential in shaping this book into its final form. I extend my sincere thanks to Nureen Syamimie Idris and Elis Irdina Idris for their roles in bringing this book to fruition, from initial drafts to final edits. Gratitude goes to Prof. Ir. Kaharudin Dimyati for the support and for sharing knowledge, which greatly influenced the content and quality of this book.

This book consists of six main chapters. Each chapter is divided into several subsections to provide detailed descriptions of the PAPR problem in wireless communication, especially MIMO-OFDM systems, a brief overview of the fundamentals of OFDM and the PAPR issue, the recent emergence of PAPR reduction techniques and its limitation, and the continuation of the PAPR problem in MIMO-OFDM systems along with an introduction to the diversity schemes of STBC, SFBC, and STFBC.

The finding of this book will rebound to the benefit of society, considering that OFDM plays an important role in wireless communication systems. The greater demand for the OFDM system justifies a more effective and efficient technique to deal with serious PAPR problems. Thus, a telecommunication company that applies the recommended approach derived from this book will be able to increase the efficiency of the radio frequency (RF) power amplifier and performance for the whole system. Furthermore, the contents of this book will help the researcher uncover critical areas that need to be focused on to develop a better solution for the high PAPR issue.

Shah Alam, Malaysia

Azlina Idris CEng
Aidatul Julia Abd Jabar
Wan Norsyafizan W. Muhamad

Contents

1	**Overview and Challenges of Wireless Communication System**	1
	1.1 Introduction	1
	1.2 Challenges of Wireless Communication	1
	1.3 MIMO-OFDM System	3
	References	4
2	**Orthogonal Frequency Division Multiplexing System and Peak-To-Average Power Ratio Issues**	7
	2.1 The Wireless—the Emergence of Wireless Communication	7
	2.2 Orthogonal Frequency Division Multiplexing (OFDM)	8
	2.3 Digital Modulation	12
	2.4 PAPR Issues in OFDM System	13
	2.4.1 Non-Linear Characteristics of HPA and DAC	14
	2.4.2 Power Saving	15
	References	15
3	**System Model of OFDM System**	19
	3.1 System Model of OFDM System	19
	3.1.1 PAPR Analysis Through OFDM Signal Equation	21
	3.1.2 Bit Error Rate (BER) Analysis Through OFDM Signal Equation	24
	References	25
4	**PAPR Reduction Techniques in OFDM System**	27
	4.1 Signal Distortion Technique	27
	4.1.1 Clipping Technique	27
	4.1.2 Coding Technique	28
	4.1.3 Multiple Signaling and Probabilistic Techniques	30
	4.1.4 Selective Mapping Technique	32
	4.1.5 Selective Codeword Shift Technique	36
	4.1.6 Precoding Technique	38
	References	41

5	**PAPR in MIMO-OFDM System**	45
	5.1 Introduction	45
	5.2 MIMO-OFDM System	46
	5.3 PAPR in MIMO-OFDM System	47
	5.4 MIMO-OFDM System for 5G Networks	50
	References	50
6	**Diversity Techniques**	53
	6.1 Introduction	53
	6.2 Diversity Reception Combining	56
	References	60
Index		63

About the Authors

Azlina Idris CEng is a Professor at the Universiti Teknologi MARA (UiTM) Shah Alam, Selangor, Malaysia. She obtained her Ph.D. in Wireless Communication from University Malaya (UM), Malaysia. She is a member of Wireless Communication Technology (WiCOT) Research Interest Group (RIG), and her research interests include OFDM/OFDMA transmission, MIMO technology, Internet of Things (IoT), channel coding, interference management and mitigation, 5G system, and channel modeling. She has published around 170 articles in journals and paper proceedings. She has examined more than 23 Ph.D./Master by research and Mix Mode students as external examiner, locally and abroad. e-mail: azlina831@uitm.edu.my

Aidatul Julia Abd Jabar is a Ph.D. candidate at Faculty of Electrical Engineering, Universiti Teknologi MARA (UiTM) Shah Alam, Selangor. She was graduated her first degree from Universiti Teknologi Malaysia (UTM) in Bachelor of Electrical Engineering majoring in Telecommunication. She has received the Master of Engineering in Telecommunication from Universiti Malaya (UM). Her research interests mainly focus on wireless communications, OFDM/F-OFDM systems, and power consumption reduction in wireless communications devices. e-mail: juliajabar@gmail.com

Dr. Wan Norsyafizan W. Muhamad received her Bachelor in Electrical Engineering from the Universiti Malaya, Malaysia, in 2002. She completed her Master of Electrical Engineering from Universiti Malaya, Malaysia, in 2009. She obtained her Ph.D. in 2017 from University of Newcastle, Australia. Currently, she is a Senior Lecturer in Faculty of Electrical Engineering at Universiti Teknologi MARA, Shah Alam, Malaysia. Her current research interests are in the area of wireless communication (Physical and MAC cross layer optimization). e-mail: syafizan@uitm.edu.my

Abbreviations

3GPP	3rd Generation Partnership Project
ADSL	Asymmetric Digital Subscriber Line
BER	Bit Error Rate
BLAST	Bell Laboratories Layer Space-Time
BPSK	Binary Phase Shift Keying
BS	Base Stations
BWA	Broadband Wireless Access
CCDF	Complementary Cumulative Distribution Function
CF	Crest Factor
CP	Cyclic Prefix
CSPS	Cyclically Shifted Phase Sequences
DAC	Digital-to-Analog Converter
DCT	Discrete Cosine Transform
DFT	Discrete Fourier Transform
DHT	Discrete Hartley Transform
DOCSIS	Data Over Cable Service Interface Specification
DPP	Data-Position Permutation
DVB	Digital Video Broadcasting
EARTH	Energy Aware Radio and Network Technologies
EGC	Equal Gain Combining
FDM	Frequency Division Multiplexing
FEC	Forwarding Error Check
FFT	Fast Fourier Transform
HPA	High-Power Amplifier
HSLM	Hadamard-based SLM
HSPA	High-Speed Packet data Access
IBO	Input Back-Off
ICI	Inter-Channel Interference
IFFT	Inverse Fast Fourier Transform
ISI	Inter-Symbol Interference
KSP	Known Symbol Padding

LDPC	Low-Density Parity Check
LTE	Long-Term Evolution
MBWA	Mobile Broadband Wireless Access
MC	μ-law Companding
MIMO	Multiple Input Multiple Output
MISO	Multiple Input Single Output
ML	Maximum Likelihood
MoCA	Multimedia over Coax Alliance
MRC	Maximal Ratio Combining
OFDM	Orthogonal Frequency Division Multiplexing
OFDMA	Orthogonal Frequency Division Multiple Access
PAN	Personal Area Network
PAPR	Peak-to-Average Power Ratio
PLC	Power Line Communication
PTS	Partial Transmit Sequences
QAM	Quadrature Amplitude Modulation
QC-LDPC	Quasi-Cyclic LDPC
QoS	Quality of Service
QPSK	Quadrature Phase Shift Keying
RMC	Root-based μ-law Companding
SC	Selection Combining
SC-FDMA	Single-Carrier Frequency Division Multiple Access
SCS	Scrambling Codeword Shifting
SFBC	Space-Frequency Block Code
SI	Side Information
SIMO	Single Input Multiple Output
SISO	Single Input Single Output
SLM	Selective Mapping
SM	Spatial Multiplexing technique
SNR	Signal-to-Noise Ratio
STBC	Space-Time Block Code
STC	Space-Time Code
STF	Space-Time-Frequency
STFBC	Space-Time-Frequency Block Code
UWB	Ultra-Wideband
V-BLAST	Vertical Bell Laboratories Layered Space-Time
VDSL	Very high-speed Digital Subscriber Line
WCDMA	Wideband Code Division Multiple Access
WHT	Walsh-Hadamard Transform
WiMAX	Worldwide Interoperability for Microwave Access
WLAN	Wireless LAN
ZP	Zero Padding

Symbols

f_0	Carrier frequency/initial frequency
f_k	Subcarrier frequencies
Δ_f	Frequency spacing
$X(t)$	QAM signal
X_k	OFDM symbol sequence
η	PA efficiency
$s(t)$	Modulated subcarriers
$z(t)$	Received signal
$Z(k)$	Demodulated subcarrier
$h(t)$	Time-varying channel impulse response
$n(t)$	Noise waveform
PAPR	Peak-to-average power ratio
E	Average signal power
$F(\delta)$	Cumulative Rayleigh distribution
BER	Bit error rate
$\frac{E_b}{N_0}$	Energy per bit to noise power spectral density ratio
Z_S	Output of symbol sequences

List of Figures

Fig. 2.1	Typical power consumption distribution in radio access technology (adapted from [2, 4])	8
Fig. 2.2	Comparison of the bandwidth utilization for FDM and OFDM (adapted from [12])	9
Fig. 2.3	Block diagram of OFDM transmission system (adapted from [13])	10
Fig. 2.4	Block diagram of QAM modulator (adapted from [23])	12
Fig. 2.5	Input and output characteristic of an HPA (adapted from [16])	14
Fig. 3.1	OFDM modulation of inverse fast fourier transform (IFFT) (adapted from [1])	20
Fig. 3.2	Signal in channel model (adapted from [2])	20
Fig. 3.3	OFDM signal demodulation at receiver using FFT (adapted from [2])	21
Fig. 3.4	Received signal using demodulation process of OFDM	22
Fig. 4.1	Block diagram of conventional SLM technique (adapted from [40])	33
Fig. 4.2	Block diagram of SCS technique in OFDM system (adapted from [1])	37
Fig. 5.1	Comparison of diversity and spatial multiplexing technique	46
Fig. 5.2	MIMO system (adapted from [3])	47
Fig. 5.3	Block diagram of individual SLM technique (adapted from [9])	48
Fig. 5.4	Block diagram of concurrent SLM technique (adapted from [11])	49
Fig. 6.1	STBC encoding using Alamouti scheme	54
Fig. 6.2	SFBC encoding using Alamouti scheme	55
Fig. 6.3	STFBC encoding using Alamouti scheme	56
Fig. 6.4	Diversity reception combining (adapted from [12])	58

List of Tables

Table 2.1	Symbol rate comparison for different modulation format [22]	12
Table 4.1	SCS bit arrangement for different number of shift factors using 64-QAM [1]	38
Table 4.2	Entries of different precoding matrices [64]	39

Chapter 1
Overview and Challenges of Wireless Communication System

1.1 Introduction

The wireless communication systems has made significant progress since its inception, revolutionizing the way people communicate. A great leap in communication systems was observed when wireless communication was introduced, in which the world witnessed the transition from wired to wireless operation. With this technological advancement, users worldwide can be connected through short- and long-range communication. The rapid development of wireless communications has contributed to the enormous growth in wireless users over the past few decades. With the recent lifestyle of modern society, high data rate wireless applications such as Internet access, multimedia stream, and mobile computing have become inseparable from the daily lives of human beings. However, to provide users with high-speed, reliable wireless communication experience, several technical challenges such as limited bandwidth and signal fading inherent to wireless channels need to be resolved.

1.2 Challenges of Wireless Communication

These challenges could be overcome through a practical solution of MIMO and OFDM implementation. MIMO is a multiple antenna system invented to improve signal propagation for communication over the wireless channel by exploiting multipath propagation compared to the traditional single-antenna systems. By employing multiple transmit and receive antennas, the adverse effects of the wireless propagation environment can be significantly reduced. On the other hand, OFDM is perfect for a high data rate wireless system as a multicarrier modulation with high spectral efficiency. The concept of OFDM is to divide the radio channel into multiple narrowband, low-rate, overlapped sub-channels or subcarriers to allows multiple symbols to be transmitted parallelly while maintaining total bandwidth similar to

© The Author(s), under exclusive license to Springer Nature Singapore Pte Ltd. 2025
A. I. CEng et al., *MIMO-OFDM Systems with Diversity Technique*,
SpringerBriefs in Applied Sciences and Technology,
https://doi.org/10.1007/978-981-96-1001-3_1

conventional single-carrier modulation. On top of that, OFDM is also equipped with a unique feature that is immunity against frequency selective fading compared to single-carrier systems [1]. Therefore, it makes sense why OFDM is adopted in most wireless broadband standards, including Third Generation Partnership Project Long Term Evolution (3GPP-LTE) and Worldwide Interoperability for Microwave Access (WiMAX) [2, 3].

Nowadays, environmentally friendly technologies or green technologies have been a priority in developing new technology as it reduces the impact of humans on the environment. So far, research efforts in wireless communication direct attention to spectrum efficiency, transmission reliability, data rate, and services provided to users. However, most recent research efforts have disregarded the implication of the wireless network's environmental responsibility in energy efficiency and environmental impact [4]. In response to this issue, green communications have been promoted by both academia and industry as a new solution for reducing energy consumption or improving energy efficiency while meeting the requirement of high quality of service (QoS) and spectral efficiency [5–8]. A study reported that wireless communication systems are responsible for at least 0.5% of the global energy consumption [9] where base station (BS) operation contributes to the highest energy consumption for this industry [10]. This is because transmitted signals from the transmitter in wireless access networks consume more energy than traditional fixed-line networks where physical wires connect the communicating nodes [11]. Before any transmission in the wireless medium is performed, the transmitted signal must be amplified to avoid the attenuation effects resulting from path loss, random fading, and interferences. This power-consuming process is performed by a high-power amplifier (HPA), contributing to approximately 50–80% of power consumption distribution in the wireless system [12]. Therefore, the HPA efficiency seems to be the main component for achieving low power consumption [13]. On the other hand, the high-power consumption of HPA is caused by the non-linearity and inefficiency characteristics of the amplifier itself. The non-linear amplification due to the signal processing leads to significant performance degradation, while inefficient amplification associated with HPA architecture causes severe reduction of system energy efficiency [13].

One of the alternatives to achieved HPA efficiency is through a low complex approach associated with signal design. In the OFDM system, the inevitable high PAPR issue happens to be caused by the signal modulation [14] process. According to Correia et al. [7], the modulation schemes used in Wideband Code Division Multiple Access (WCDMA) or High-Speed Packet data Access (HSPA) and Long-Term Evolution (LTE) have caused the PAPR to exceed 10 dB. This PAPR problem is more critical in the OFDM system compared with single-carrier systems. The occurrence of high PAPR in the OFDM system is due to the nature of the modulation itself, where transmitted signals are formed through a summation of multiple subcarriers [15, 16]. High PAPR is known to be one of the major limitations in implementing OFDM systems [17], and this problem exists in MIMO-OFDM systems [18, 19]. Apart from power efficiency [20, 21], high PAPR could also lead to a chain of problems such as the increased probability of energy spilling to the adjacent channel, signal distortion in the power amplifier, and error rate performance degradation due to the system constraint to a limited peak power of dynamic range in transmitter amplifier. For this reason, high PAPR reduction has become one of the active research topics in the wireless communication field.

Many techniques have been proposed to address the serious PAPR problem in OFDM systems. However, the problem with these techniques is that their effects on other parameters are not always positive. These effects include a reduction in data rate, a decrease in BER, and increased complexity or requirement of side information (SI). Therefore, multiple signaling and probabilistic techniques such as interleaving, selective mapping (SLM), and precoding have received great attention and attracted scientists' interest due to their PAPR reduction performance compared with other existing techniques [22]. The fundamental concept of the multiple signaling technique is to generate a permutation of the multicarrier signal by choosing the signal with the minimum PAPR for transmission.

In contrast, probabilistic techniques modify different parameters in the OFDM signal, optimizing them to minimize the PAPR. However, this technique has a theoretical limitation of high computational complexity due to the additional process added to the system and BER degradation related to the side information issue [23]. Realizing the significance of this issue, a new interleaving technique is proposed in this study to address the problem mentioned above.

1.3 MIMO-OFDM System

As one of the most promising techniques for broadband wireless access schemes, MIMO-OFDM is still unable to escape the high PAPR problem inherited from the OFDM system. Vertical Bell Laboratories Layered Space–Time (V-BLAST) is a MIMO-OFDM spatial multiplexing scheme that exploits the multipath channel in a highly scattering wireless environment to provide an increase in data rate through simultaneous transmission over multiple transmission antennas. This scheme possesses a significant multiplexing gain in exchange for its lack of diversity gain. This implies that each antenna is independently similar to the single-input single-output (SISO) case and, therefore, shares the same PAPR value [24, 25]. A joint technique using a diversity scheme has become one of the major focuses in the research community to achieve maximum diversity gain through the improvement of BER performance at the receiver. Formulation of permutation in the interleaving technique is not the only factor contributing to the good PAPR reduction. Another aspect, such as the codeword structure itself, can also be manipulated for that purpose. However, the relationship between codeword structure and PAPR reduction capability has not yet been studied, so it is not obvious which codeword structure will give good PAPR performance, whether altered or unaltered codeword structure. Therefore, this book aims to overcome the high PAPR faced by the OFDM systems that takes advantage of data permutation and codeword structure manipulation. This technique will have low computational complexity and is capable of preventing BER performance degradation. Besides, a diversity scheme MIMO-OFDM with different diversity properties such as space–time, space-frequency, and space–time-frequency is also presented as a solution for improving PAPR and BER performance in the MIMO-OFDM system. The results have demonstrated that this technique was effective in serving its purpose in both OFDM and MIMO-OFDM systems.

References

1. T. Sultana, R.A. Akhi, J.H. Turag, S. Najeeb, Study of different candidates of modulation schemes for 5G communication systems, in *3rd International Conference on Electrical, Computer and Communication Engineering, ECCE 2023* (2023), pp. 1–5. https://doi.org/10.1109/ECCE57851.2023.10101611
2. F.M. Mustafa, H.M. Bierk, M.N. Hussain, Reduction of PAPR pattern with low complexity using hybrid-Pts scheme for the 4G and 5G multicarrier systems. J. Eng. Sci. Technol. **16**(4), 3481–3504 (2021)
3. E. Hossain, Editorial: fourth quarter 2015 IEEE communications surveys and tutorials. IEEE Commun. Surv. Tutor. **17**(4), 1801–1805 (2015). https://doi.org/10.1109/COMST.2015.2495008
4. L.K. Chhaya, Green wireless communication. J. Telecommun. Syst. Manage. **1**(3), 1–5 (2012). https://doi.org/10.4172/2167-0919.1000104
5. T. Chen, C. Wang, Editorial green technologies for wireless communications and mobile computing. IET Commun. **5**(18), 2595–2597 (2011). https://doi.org/10.1049/iet-com.2011.0852
6. J. Joung, C.K. Ho, K. Adachi, A survey on power-amplifier-centric techniques for spectrum and energy efficient wireless communications. IEEE Commun. Surv. Tutor. **17**(1), 315–333 (2015). https://doi.org/10.1109/COMST.2014.2350018
7. L.M. Correia, D. Zeller, O. Blume, D. Ferling, Challenges and enabling technologies for energy aware mobile radio networks. IEEE Commun. Mag. 66–72 (2010). https://doi.org/10.1109/MCOM.2010.5621969
8. R. Ayre, K. Hinton, R.S. Tucker, Energy consumption in wired and wireless access networks. IEEE Commun. Mag. **49**(6), 70–77 (2011). https://doi.org/10.1109/MCOM.2011.5783987
9. G. Fettweis, E. Zimmermann, ICT energy consumption—trends and challenges, in *International Symposium on Wireless Personal Multimedia Communications (WPMC 2008) ICT* (2008), pp 2006–2009
10. A. Kumar, K. Singh, D. Bhattacharya, Green communication and wireless networking, in *International Conference on Green Computing, Communication and Conservation of Energy (ICGCE)* (2014), pp. 49–52
11. W. Vereecken, W. Van Heddeghem, M. Deruyck, B. Puype, B. Lannoo, W. Joseph, Power consumption in telecommunication networks: overview and reduction strategies. IEEE Commun. Mag. **49**(6), 62–69 (2011). https://doi.org/10.1109/MCOM.2011.5783986
12. M. Gruber, O. Blume, D. Ferling, D. Zeller, M.A. Imran, E.C. Strinati, EARTH—energy aware radio and network technologies, in *IEEE International Symposium on Personal, Indoor and Mobile Radio Communications, PIMRC* (2009), pp. 1–5. https://doi.org/10.1109/PIMRC.2009.5449938
13. T. Jiang, C. Li, C. Ni, Effect of PAPR reduction on spectrum and energy efficiencies in OFDM systems with class—a HPA over AWGN channel. IEEE Trans. Broadcast. **59**(3), 513–519 (2013). https://doi.org/10.1109/TBC.2013.2253814
14. J. Joung, C.K. Ho, S. Sun, Green wireless communications: a power amplifier perspective, in *Conference Handbook—Asia-Pacific Signal and Information Processing Association Annual Summit and Conference, APSIPA ASC 2012* (2012)

References

15. G. Berardinelli, L.Á.M.R. De Temiño, S. Frattasi, M.I. Rahman, P. Mogensen, OFDMA vs. SC-FDMA performance: comparison in local area IMT-a scenarios. IEEE Wirel. Commun. **15**(5), 64–72 (2008)
16. S. Wang, J. Sie, C. Li, Y. Chen, A low-complexity PAPR reduction scheme for OFDMA uplink systems. IEEE Trans. Wirel. Commun. **10**(4), 1242–1251 (2011)
17. R. Singh, G.K. Soni, R. Jain, A. Sharma, N.V. Tawania, PAPR reduction for OFDM communication system based on ZCT-pre-coding scheme, in *Proceedings of the 2nd International Conference on Electronics and Sustainable Communication Systems, ICESC 2021*, no. 1 (2021), pp. 555–558. https://doi.org/10.1109/ICESC51422.2021.9532776
18. A. Azeez, S. Tarannum, Performance analysis of MIMO-OFDM wireless systems using precoding and companding techniques, in *2021 7th International Conference on Advanced Computing and Communication Systems, ICACCS 2021*, no. 2 (2021), pp. 120–23. https://doi.org/10.1109/ICACCS51430.2021.9441760
19. J.P. Yadav, R. Mishra, An analysis of PAPR reduction from the point of view of BER performance in next-generation MIMO-OFDM wireless systems. Int. J. Res. Develop. Appl. Sci. Eng. (IJRDASE) **24**(1) (2024)
20. M. Hu, W. Wang, W. Cheng, H. Zhang, Initial probability adaptation enhanced cross-entropy-based tone injection scheme for PAPR reduction in OFDM systems. IEEE Trans. Veh. Technol. **70**(7), 6674–6683 (2021). https://doi.org/10.1109/TVT.2021.3078736
21. X. Cui, K. Liu, Y. Liu, Novel linear companding transform design based on linear curve fitting for PAPR reduction in OFDM systems. IEEE Commun. Lett. **25**(11), 3604–3608 (2021). https://doi.org/10.1109/LCOMM.2021.3107410
22. D.W. Lim, S.J. Heo, J.S. No, An overview of peak-to-average power ratio reduction schemes for OFDM signals. J. Commun. Netw. **11**(3), 229–239 (2009). https://doi.org/10.1109/JCN.2009.6391327
23. T. Jiang, Y. Wu, An overview: peak-to-average power ratio reduction techniques for OFDM signals. IEEE Trans. Broadcast. **54**(2), 257–268 (2008). https://doi.org/10.1109/TBC.2008.915770
24. P. Wolniansky, *V-BLAST: An Architecture for Realizing Very High Data Rates Over the Rich-Scattering Wireless Channel* (2014). https://doi.org/10.1109/ISSSE.1998.738086
25. T. Tsiligkaridis, D.L. Jones, PAPR reduction performance by active constellation extension for diversity MIMO-OFDM systems. J. Electr. Comput. Eng. **2010**(2), 2–6 (2010). https://doi.org/10.1155/2010/930368

Chapter 2
Orthogonal Frequency Division Multiplexing System and Peak-To-Average Power Ratio Issues

2.1 The Wireless—the Emergence of Wireless Communication

Wireless communications are the field of communication that is attractive, reliable, and easy to access. The advancement in wireless communication technology was highly driven by the rise in the number of mobile subscribers, multimedia applications, and data rates. Over the past few years, most previous studies have concentrated on enhancing system capacity, data rates, spectral efficiency, system reliability and security, and signal interference robustness. The aspects mentioned above can be concluded as a sophisticated and advanced demand for the current lifestyle. This demand can be satisfied in three ways: the development of signal modulation, hardware implementation, and accessibility related to the number of base stations (BS).

Multicarrier modulation technique such as the OFDM system is one of the most popular signal modulation creation efforts. Compared to a conventional single-carrier modulation, the OFDM system provides many overlapping subcarriers in the available bandwidth. A substantial improvement in data rates and spectral efficiency is given by this method. Furthermore, the MIMO system increases system capacity in hardware implementation by using multiple antennas at both transmitter and receiver. MIMO works well with OFDM, which has increased the data rates significantly.

OFDM improves data rate and spectrum efficiency, but the rise in subcarriers conversely contributes to the high peak-to-average power ratio (PAPR). The serious PAPR problem induces the non-linearity of the power amplifier, which contributes to high power consumption. In addition, the advancement of wireless technology is essential only if it is balanced with green technology. Most research efforts have neglected the wireless network's environmental responsibility, such as energy efficiency and environmental impact [1]. After Energy Aware Radio and Network Technologies (EARTH) was published in 2010 on energy efficiency in wireless networks,

Fig. 2.1 Typical power consumption distribution in radio access technology (adapted from [2, 4])

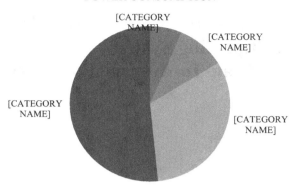

it has drawn a great deal of interest from academia and industry [2]. Different components of BS are known as the major contributors to total energy consumption [3]. The typical distribution of power consumption for radio access equipment recorded by Alcatel Lucent and Vodafone in 2009 is shown in Fig. 2.1 [4]. It indicates that the power amplifier consumed a large proportion of energy usage, approximately 50% to 80%.

It is well known that a large portion of the overall energy consumption is absorbed by the high-power amplifier (HPA) in wireless communication devices. While there are urges to reduce the energy consumption of HPA, this effort needs to consider the requirement for high linearity of HPA to achieve the quality of transmitted signal specified by the standards. Hence, the HPA must perform well below saturation, where the power consumption would be considerably high at this stage. Unfortunately, this requirement has resulted in low energy efficiency. The main problem here is that signals passing through HPA are varied and convey high power, causing high PAPR to emerge in OFDM systems. The modulation scheme used in WCDMA/HSPA and LTE, for example, is characterized by highly varied signal envelopes with a PAPR greater than 10 dB [2]. In this case, the PAPR reduction solution must be applied to improve the power efficiency and, at the same time, prevent the HPA process from increasing its linearity range. Besides, other effects such as signal distortion that increases BER and energy spill to the adjacent channel that interferes with neighboring frequency have become the reason to minimize high PAPR in the OFDM system [5–7].

2.2 Orthogonal Frequency Division Multiplexing (OFDM)

OFDM is one of the most dominant cutting-edge technologies and has gained considerable attention in wireless communication. The combination of high data capacity, high spectral efficiency, low complexity, and resilience to inter-symbol interference

2.2 Orthogonal Frequency Division Multiplexing (OFDM)

(ISI) and its resistance to interference due to multipath effects [8, 9] implies an ideal platform for the current future wireless data communication transmissions. Conventionally, data transmission over radio channels is made in series, one bit after another using a single channel. Unfortunately, this is not an effective method as the channel is vulnerable to any interference, jeopardizing the whole transmission. Therefore, a different approach is adopted by OFDM. The fundamental operation of OFDM is to divide the overall wide signal bandwidth into many narrowband sub-channels. Within the overall narrowband sub-channels, the data is transmitted parallelly across the various carriers to maintain high data transmission. By doing so, the appearance of ISI can be prevented due to the increment of symbol duration relative to the channel delay spread. Furthermore, the orthogonality feature possessed by OFDM allows multiple subcarriers to overlap with each other and at the same time avoid the crosstalk incident. In addition, the carrier's orthogonal properties will provide a bandwidth saving where the reserved space can be used by more carriers [8, 10, 11].

Figure 2.2 shows the spectral efficiency of OFDM compared to conventional Frequency Division Multiplexing (FDM) as a result of orthogonality. Maintaining the orthogonality of multiple subcarriers is the main aspect in OFDM, which can be attained by carefully selecting subcarrier spacing, such as allowing the subcarrier spacing, Δf, to be equal to the reciprocal of the used symbol period, T_u.

A complete end-to-end IFFT-based OFDM system comprises a transmitter and a receiver implemented by combining different blocks as shown in Fig. 2.3. In a digitally implemented OFDM system, a serial stream of data in binary form that enters the system is directly converted into N parallel data streams through the serial-to-parallel converter where N is the number of subcarriers in the system. The parallel data each consisting of m bits are fed to digital modulation block to form a

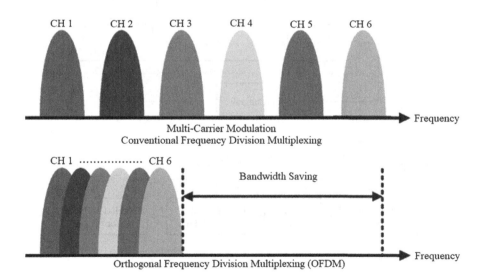

Fig. 2.2 Comparison of the bandwidth utilization for FDM and OFDM (adapted from [12])

complex number called OFDM symbols. The symbols are allocated to each of the subcarriers which are mapped to a subcarrier amplitude and phase accordance to the modulation technique used such as Binary Phase Shift Keying (BPSK), Quadrature Phase Shift Keying (QPSK), or Quadrature Amplitude Modulation (QAM). Then, inverse fast Fourier transform (IFFT) takes in the array of modulated symbols and converts it to time-domain OFDM signals. The resulting OFDM signal containing N data symbols is then converted back into serial form. Moreover, the guard interval is inserted between symbols to prevent ISI induced by multipath distortion. Next, the digital OFDM signal is converted into an analog signal by undergoing a digital-to-analog conversion process. A high-power amplifier then amplifies the baseband signal before being transmitted via a wireless channel.

Theoretically, each subcarrier can be assigned to carry part of the input data stream, but practically that is not how it works in the OFDM system. For instance, 72 out of 128 subcarriers are allocated to carry the original information data stream for the lowest rate of 3GPP LTE standard, while other subcarriers are reserved for guards and pilots. Guard subcarriers are inactive or null, and their presence is to protect against inter-channel interference (ICI) with the adjacent channel. Meanwhile, the pilot subcarrier does not carry any information taken from the input data stream. But instead, it carries some information regarding the amplitude, phase, and timing for channel estimation and synchronization between transmitter and receiver.

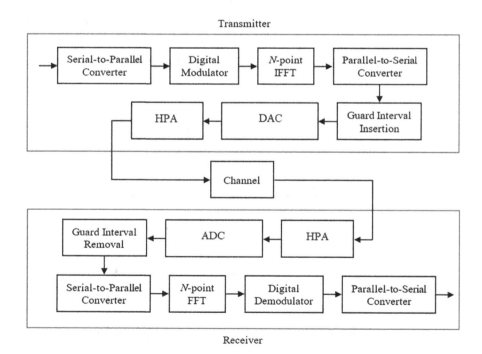

Fig. 2.3 Block diagram of OFDM transmission system (adapted from [13])

2.2 Orthogonal Frequency Division Multiplexing (OFDM)

Guard subcarrier or guard band is referred to the unused part of the OFDM symbol structure that served as a separator between successive symbols. The guard subcarrier application is crucial to prevent linear distortion such as multipath fading resulting from energy spilling into the adjacent channels. The selection of an appropriate guard interval should be well planned as it significantly impacts the system performance. The system capacity will reduce more than necessary if too large periods are chosen, whereas choosing too short ones increases the probability of interference and, thus, the error probability. The insertion of guard interval is usually less than $T_u/4$. There are several types of guard interval that can be utilized by OFDM such as zero padding (ZP) and known symbol padding (KSP), but the most commonly used is the cyclic prefix (CP). The popularity of CP is driven by its ability to maintain the receiver carrier synchronization. The OFDM symbol duration, T_s, can be computed as [14]

$$T_s = T_u + T_g \tag{2.1}$$

where T_u is used symbol duration and T_g is guard time.

At the receiver, the process is reversed. Firstly, the HPA receives the incoming signal and converts it to a baseband signal. Next, the analog-to-digital converter converts the baseband signal to digital form and passes it on to the fast Fourier transform (FFT) block. The FFT block transforms the time domain signal back to the array of subcarriers before the demodulation process takes place in the frequency domain. Digital demodulator reproduces the bit stream from each subcarrier by exploiting the amplitude and phase of the subcarriers.

OFDM is a modulation format that is used widely for many of the latest wireless and telecommunications standards worldwide due to its advantages. Several wired communication standards such as Asymmetric Digital Subscriber Line (ADSL) [15], Very high-speed Digital Subscriber Line (VDSL), Digital Video Broadcasting (DVB) [16, 17], Power Line Communication (PLC), Multimedia over Coax Alliance (MoCA), and Data Over Cable Service Interface Specification (DOCSIS) specifically adopt OFDM techniques.

Not limited to wired applications, OFDM has also been widely adopted in numerous high data rate wireless standards such as wireless LAN (WLAN) radio interfaces IEEE 802.11a [18], g, n, ac, ah, and HIPERLAN/2, DAB+ [19], and Digital Radio Mondiale and Wireless Personal Area Network (PAN) ultra-wideband (UWB) IEEE 802.15.3a. The OFDM-based multiple access technology OFDMA is also used in several 4G and pre-4G cellular networks and mobile broadband standards like the wireless MAN/broadband wireless access (BWA) standard IEEE 802.16e, also known as Mobile WiMAX [20], the mobile broadband wireless access (MBWA) standard IEEE 802.20, the downlink of the 3GPP Long Term Evolution (LTE) fourth-generation mobile broadband standard, and WLAN IEEE 802.11ax [21].

2.3 Digital Modulation

Modulation has become a fundamental element to all wireless communications, including the OFDM system. Nowadays, most wireless transmissions are made digitally. With the limited bandwidth available, the selection of modulation techniques is more critical than ever. Furthermore, digital modulation contributes to the system's increased capacity due to its capability to convey more bits of data per symbol compared to analog ones. Therefore, higher modulation orders such as QAM can offer a much faster data rate. Table 2.1 summarizes the bit rate and symbol rate for a different type of modulation technique.

Figure 2.4 depicts the basic block diagram of the QAM modulator. The generation of the QAM signal begins by splitting the data bit stream into two equal parts. Then, one data signal is multiplied by a cosine while the other data signal is multiplied by a sine to create an in-phase (I) signal and quadrature (Q) signal. As such, the two carrier waves of the same frequency are 90° out of phase with each other. Finally, the two resultant signals are summed together to produce the QAM signal.

The QAM signal, (t), can be expressed as [23]

$$X(t) = \Re\{[I(t) + iQ(t)] \cdot e^{i2\pi f_0 t}\} = I(t)cos(2\pi f_0 t) - Q(t)sin(2\pi f_0 t) \qquad (2.2)$$

Table 2.1 Symbol rate comparison for different modulation format [22]

Modulation scheme	Bits per symbol, m	Symbol rate
BPSK	1	1 × bit rate
QPSK	2	1/2 bit rate
8 PSK	3	1/3 bit rate
16 QAM	4	1/4 bit rate
32 QAM	5	1/5 bit rate
64 QAM	6	1/6 bit rate
128 QAM	7	1/7 bit rate

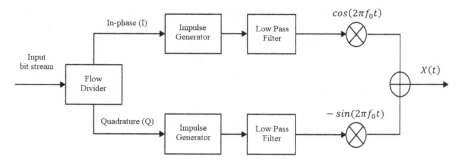

Fig. 2.4 Block diagram of QAM modulator (adapted from [23])

where $i^2 = 1$, $I(t)$ and $Q(t)$ are the modulating signals, f_0 is the carrier frequency, and $\Re(t)$ is the real part. Each symbol power can be calculated using [24]

$$X_{power} = \frac{|X(t)|^2}{2} \quad (2.3)$$

Subsequently, the OFDM symbol sequence can be written as

$$X_k = [X_0, X_1, X_2, \ldots, X_{N-1}] \, 0 \leq k \leq N - 1 \quad (2.4)$$

where N represents the number of subcarriers in the symbol duration T_u.

2.4 PAPR Issues in OFDM System

Energy efficiency is one of the necessary features for future mobile communications networks. This factor is driven by the increasing energy cost of network operations by 50% of today's total operational cost. The effective way to reduce energy costs is to increase the efficiency of the high-power amplifier (HPA). However, the efficiency of the HPA is directly related to the peak-to-average power ratio (PAPR) of the input signal. High PAPR is one of the major issues encountered by the OFDM system. The nature of the modulation itself in which all the subcarrier components are added together via IFFT operation has been the cause of this problem. Therefore, the HPA should be operated near the saturation region to achieve maximum efficiency. However, this does not happen due to the high PAPR in the OFDM signal. Still, instead, the HPA will cross over to the non-linear region, causing immense in-band distortions, adjacent channel interference, along spectrum re-growth in the transmitted signal. Therefore, operating HPA in the linear region with a huge back-off is preferable to minimize the non-linear effects, as shown in Fig. 2.5. Unfortunately, this action comes at a price, the efficiency has to be compromised, and, at the same time, the performance of the OFDM system will be degraded.

The downside of PAPR has a different effect on both uplink and downlink channels. By implementing a broad range of power amplifiers and complicated PAPR reduction techniques, the excessive PAPR in the downlink transmission can be easily resolved. Still, in return, the costs will increase. However, the same solution is not suitable for implementing in the uplink channel due to the limitation in low processing power devices. On the contrary, using a higher linear range of power amplifiers becomes crucial, which is costly. Therefore, Single-Carrier Frequency Division Multiple Access (SC-FDMA) has been chosen by the 3rd Generation Partnership Project (3GPP) as the Long-Term Evolution (LTE) uplink transmission due to its low PAPR characteristic, which greatly benefits in terms of transmit power efficiency and reduced cost of the power amplifier.

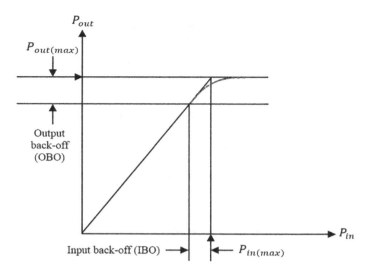

Fig. 2.5 Input and output characteristic of an HPA (adapted from [16])

2.4.1 Non-Linear Characteristics of HPA and DAC

Most radio systems use the HPA in the transmitter to gain adequate transmission power. For the HPA to achieve maximum output power efficiency, it must be operated at or near the saturation region. Moreover, the HPA's non-linear characteristic is highly sensitive to the variation in signal amplitudes. In OFDM systems, the transmitted signal can have a high PAPR resulting from the amplitude of several subcarriers having high peaks added simultaneously.

Consequently, this will cause in-band distortions and out-of-band radiation to be introduced into the systems. However, that is not all, since the HPA will suffer from additional interference that increases BER performance. Therefore, the HPA needs to be forced to operate in its linear region with a large dynamic range to alleviate this problem. However, this solution is impractical since a linear amplifier will have poor efficiency, and the implementing cost is expensive. Power efficiency is crucial in wireless communication as it offers sufficient area coverage and reduces power consumption. For that reason, the non-linear HPA needs to operate efficiently with low back-off values and come up with a possible solution to the interference issue brought about. Thus, the practical way to tackle this problem is to prevent such interference in the first place by minimizing the PAPR of the transmitted signal itself.

On the other hand, a high precision digital-to-analog converter (DAC) with a wide dynamic range is needed to accommodate the large peaks of the OFDM signals, which is not a viable solution as the implementation cost is quite expensive. As an alternative, a low precision DAC could be used as it is more cost-efficient but will have significant quantization noise. Since drastic sacrifices are needed to modifying the non-linear components to support high PAPR, it seems like reducing the large variations in the OFDM signal would be the best option instead of dealing with the non-linear devices [25].

2.4.2 Power Saving

Research has proven that a reduction in PAPR would lead to significant power savings. Power savings become more relevant when mobile terminals are in the system as these devices have limited battery life. Consider the most linear amplifiers, the Class A power amplifier which has a maximum PA efficiency (η_{max}) of 50% [26, 27]. Assuming an ideal linear model for the HPA, where the linear amplification is achieved up to the saturation point [26], thus, the PA efficiency in this amplifier is given by [25]

$$\eta = \frac{\eta_{max}}{\text{PAPR}} = \frac{0.5}{\text{PAPR}} \qquad (2.5)$$

where the PAPR is expressed in linear units. Let's go through a few examples to understand better the relationship between PAPR reduction in OFDM signal and power saving. Given an OFDM signal having 64 subcarriers and without any PAPR reduction scheme applied in the system. To ensure that the probability of the clipped OFDM frames is kept below 0.01%, the system required an input back-off (IBO) equivalent to PAPR (dB) = 11.4 dB ($\approx$13.80), which is the PAPR at the 10^{-4} probability level. Thus, the calculated efficiency of HPA is $\eta = 0.5/13.80 \approx 3.6\%$. Now, a PAPR reduction technique is applied to the same system and 3 dB PAPR reduction is achieved. Hence, the new required IBO is equivalent to PAPR (dB) = 8.4 dB $\approx$ 6.92 and the PA efficiency becomes $\eta = 0.5/6.92 \approx 7.23\%$, which is equivalent to twice the efficiency. Therefore, owning low power efficiency becomes a strong motivation toward the reduction of PAPR in OFDM systems.

References

1. L.K. Chhaya, Green wireless communication. J. Telecommun. Syst. Manage. **1**(3), 1–5 (2012). https://doi.org/10.4172/2167-0919.1000104
2. L.M. Correia, D. Zeller, O. Blume, D. Ferling, Challenges and enabling technologies for energy aware mobile radio networks. IEEE Commun. Mag. 66–72 (2010). https://doi.org/10.1109/MCOM.2010.5621969
3. H.L. Hung, Y.F. Huang, C.C. Wei, R.C. Chen, Performance of PTS-based firefly algorithm scheme for PAPR reduction in SFBC MIMO-OFDM communication systems, in *IEEE International Symposium on Computer, Consumer and Control, IS3C 2016* (2016), pp. 854–857. https://doi.org/10.1109/IS3C.2016.217

4. L.M.P. Larsen, H.L. Christiansen, S. Ruepp, M.S. Berger, Toward greener 5G and beyond radio access networks-a survey. IEEE Open J. Commun. Soc. **4**, 768–797 (2023). https://doi.org/10.1109/OJCOMS.2023.3257889
5. D.W. Lim, S.J. Heo, J.S. No, An overview of peak-to-average power ratio reduction schemes for OFDM signals. J. Commun. Netw. **11**(3), 229–239 (2009). https://doi.org/10.1109/JCN.2009.6391327
6. T. Jiang, Y. Wu, An overview: peak-to-average power ratio reduction techniques for OFDM signals. IEEE Trans. Broadcast. **54**(2), 257–268 (2008). https://doi.org/10.1109/TBC.2008.915770
7. L. Wang, C. Tellambura, An overview of peak-to-average power ratio reduction techniques for OFDM systems, in *IEEE International Symposium on Signal Processing and Information Technology, ISSPIT* (2006), pp. 840–845. https://doi.org/10.1109/ISSPIT.2006.270915
8. Y. Wu, W.Y. Zou, Orthogonal frequency division multiplexing: a multi-carrier modulation scheme. IEEE Trans. Consum. Electron. **41**(3), 392–399 (1995). https://doi.org/10.1109/30.468055
9. N. LaSorte, W.J. Barnes, H.H. Refai, The history of orthogonal frequency division multiplexing, *GLOBECOM—IEEE Global Telecommunications Conference* (2008), pp. 3592–3596. https://doi.org/10.1109/GLOCOM.2008.ECP.690
10. A. Sharma, A.K. Singh, Orthogonal frequency division multiplexing and its applications. Int. J. Comput. Trends Technol. **38**(1), 21–23 (2016). https://doi.org/10.14445/22312803/ijctt-v38p105
11. A. Madhusudhan, S.K. Sharma, Improved PAPR reduction with low complexity in OFDM and filtered-OFDM (5G) systems, in *International Conference on Innovative Advancement in Engineering and Technology (IAET-2020)* (2020), pp. 1–10
12. M.M. Albogame, K.M. Elleithy, Enhancement orthogonal frequency division multiplexing (OFDM) in wireless communication network by using PTS (Partial Transmit Sequences) technique, in *29th International Conference on Computers and Their Applications, CATA 2014* (2014), pp. 73–78
13. A. Nayak, A. Goen, A review on PAPR reduction techniques in OFDM system. Int. J. Adv. Res. Electr. Electron. Instrum. Eng. **5**(4), 2767–2772 (2016). https://doi.org/10.15662/IJAREEIE.2016.0504113
14. Y.S. Cho, J. Kim, W.Y. Yang, *MIMO-OFDM Wireless Communications with MATLAB* (John Wiley & Sons, 2010)
15. V.V. Panicker, A. Najia, A. Koshy, A distortion based iterative filtering and clipping for better PAPR reduction in OFDM, in *ICCISc 2021 - 2021 International Conference on Communication, Control and Information Sciences, Proceedings*, vol. 1 (2021), pp. 1–6. https://doi.org/10.1109/ICCISc52257.2021.9485004
16. A.A.A. Wahab, H. Husin, W.A.F.W. Othman, S.S.N. Alhady, F.N. Nursyadza, PAPR reduction using genetic algorithm in OFDM system. Appl. Modelling Simul. **2**(03), 89–94 (2018)
17. B. Tang, K. Qin, H. Mei, C. Chen, Iterative clipping-noise compression scheme for PAPR reduction in OFDM systems. IEEE Access **7**, 134348–134359 (2019). https://doi.org/10.1109/ACCESS.2019.2941389
18. S. Gupta, A. Goel, Generalized trapezoidal companding technique for PAPR reduction in OFDM systems. J. Opt. Communi. **42**(2), 341–349 (2021). https://doi.org/10.1515/joc-2018-0083
19. B. O'Neill, DAB Eureka-147: The European vision for digital radio, in *Pre-Conference: The Long History of New Media: Contemporary and Future Developments Contextualized, International Communication Association, Annual Convention. Montréal* (2008), pp. 1–25
20. S. Kiambi, E. Mwangi, G. Kamucha, PAPR reduction in MIMO-OFDM systems using low-complexity additive signal mixing. J. Commun. **16**(11), 468–478 (2021). https://doi.org/10.12720/jcm.16.11.468-478
21. J. Zyren, W. Mccoy, Overview of the 3GPP long term evolution physical layer. 3GPPEVOLUTIONWP (2007)

References

22. M. Razaul, S. Mushfiq, M. Asif, Comparative analysis of various wireless digital modulation techniques with different channel coding schemes under AWGN channel. Int. J. Comput. Appl. **161**(3), 30–34 (2017). https://doi.org/10.5120/ijca2017913139
23. L.E. Franks, Carrier and bit synchronization in data communication—a tutorial review. IEEE Trans. Commun. **28**(8), 1107–1121 (1980). https://doi.org/10.1109/TCOM.1980.1094775
24. E. Abdullah, A. Idris, A. Saparon, Minimizing high PAPR in OFDM system using circulant shift codeword. J. Teknol. **78**(2), 135–140 (2016) [Online]. Available: www.jurnalteknologi.utm.my
25. F. Sandoval, G. Poitau, F. Gagnon, Hybrid peak-to-average power ratio reduction techniques: review and performance comparison. IEEE Access **5**, 27145–27161 (2017). https://doi.org/10.1109/ACCESS.2017.2775859
26. J. Wang et al., Spectral efficiency improvement with 5G technologies: results from field tests. IEEE J. Sel. Areas Commun. **35**(8), 1867–1875 (2017). https://doi.org/10.1109/JSAC.2017.2713498
27. K. Anoh, C. Tanriover, M.V. Ribeiro, B. Adebisi, C.H. See, On the fast DHT precoding of OFDM signals over frequency-selective fading channels for wireless applications. Electronics (Switzerland) **11**(19), 1–20 (2022). https://doi.org/10.3390/electronics11193099

Chapter 3
System Model of OFDM System

3.1 System Model of OFDM System

A model of the OFDM system that was used in this research is discussed in this section. The OFDM system transmits a signal using several subcarriers to transmit several data streams simultaneously. Using IFFT, which consists of N OFDM complex modulators, $e^{j2\pi fkt}$, these multiple subcarriers are modulated. One subcarrier corresponds to each OFDM modulator. An array of parallel signals can be generated by combining evenly spaced N subcarriers as shown in Fig. 3.1.

The subcarrier frequencies, $f_k = f_0 + k\Delta f$, in OFDM are selected in such a way that the subcarriers are orthogonal to each other where f_0 is the initial frequency, k is the number of modulated subcarriers, and Δf is the frequency spacing $1/T$. Mathematically, the OFDM baseband signal in the time domain with the use of any common modulation scheme such as phase shift keying (PSK) or QAM can be represented as

$$s(t) = \frac{1}{\sqrt{N}} \sum_{k=0}^{N-1} X_k \cdot e^{j2\pi f_k t} \quad 0 < t < NT \tag{3.1}$$

where $s(t)$ is the sum of modulated subcarriers, N is the number of subcarriers, and $X_k = X_0, X_1, X_2, \ldots, X_{N-1}$ are the modulation symbols. The number of bits per symbol, m, is also determined by the selection of the modulation scheme. The summation of subcarrier, N, produces a total of NT period of OFDM signal. The transmission of the OFDM signal from the transmitter to the receiver is done through a wireless channel.

Hence, the received signal can be expressed as

$$z(t) = h(t) \cdot s(t) + n(t) \tag{3.2}$$

© The Author(s), under exclusive license to Springer Nature Singapore Pte Ltd. 2025
A. I. CEng et al., *MIMO-OFDM Systems with Diversity Technique*,
SpringerBriefs in Applied Sciences and Technology,
https://doi.org/10.1007/978-981-96-1001-3_3

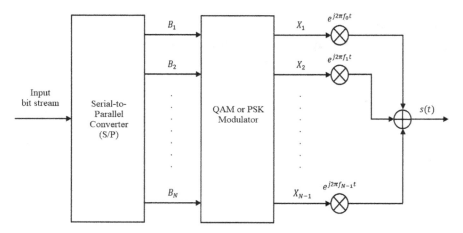

Fig. 3.1 OFDM modulation of inverse fast fourier transform (IFFT) (adapted from [1])

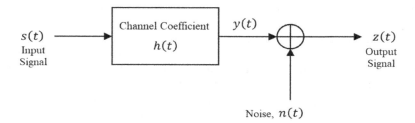

Fig. 3.2 Signal in channel model (adapted from [2])

where $s(t)$ is the input signal, $h(t)$ are the time-varying channel impulse response, and $n(t)$ is the noise waveform. All parameters contained in the channel model can be described as shown in Fig. 3.2.

This study considers only time-variant channels for simplicity purposes, and the demodulated signal at the receiver can be expressed as [2]

$$z(t) = \int h(\tau) \cdot s(t-\tau) dt + n(t) \tag{3.3}$$

The FFT block which consists of N complex demodulators will be in charge of the demodulation process for each subcarrier and is transferred to a bank of integrators afterwards as illustrated in Fig. 3.3.

Thus, the demodulated subcarrier can be represented by [2]

$$Z(k) = \frac{1}{T}\int_0^T z(t).e^{-j2\pi f_k t} dt k = 0, 1, 2, \ldots, N-1 \tag{3.4}$$

3.1 System Model of OFDM System

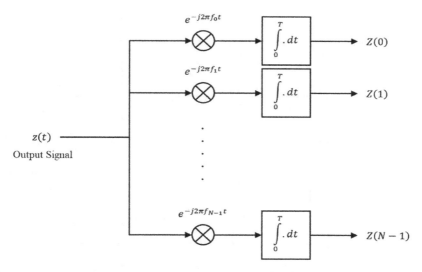

Fig. 3.3 OFDM signal demodulation at receiver using FFT (adapted from [2])

Figure 3.4 shows the OFDM signal in frequency domain consisting of a discrete signal for OFDM symbols. The multicarrier signal using OFDM system is written as

$$Z(t) = \sum [Z(0) + Z(1) + \cdots + Z(N-1)] \qquad (3.5)$$

Equation 3.5 expresses as $Z(t)$ the summation of demodulated signals, where $Z(0), Z(1), \ldots, Z(N-1)$ are received signals respectively.

OFDM signal demodulation is a process that reverses the modulation process, allowing the original data stream to be recovered at the receiver. It uses FFT for frequency domain analysis and channel coefficients to represent the channel's impact on the transmitted signal. This process is crucial in communication systems as it enables the receiver to interpret the transmitted data.

3.1.1 PAPR Analysis Through OFDM Signal Equation

OFDM systems composed of a huge number of independently modulated subcarriers can instantaneously cause high peak values in the time domain compared to single-carrier systems. The summation of subcarriers modulated by IFFT can be represented as [3]

$$s(t) = \frac{1}{\sqrt{N}} \sum_{k=0}^{N-1} X_k . e^{j2\pi f_k t} \quad 0 < t < NT \qquad (3.6)$$

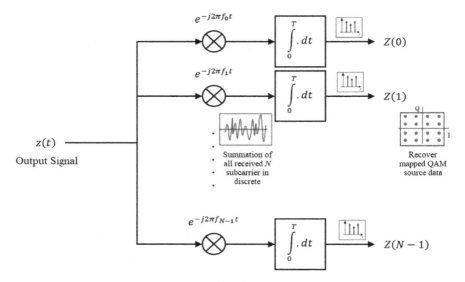

Fig. 3.4 Received signal using demodulation process of OFDM

where X_k is the modulated symbol with the frequency of f_k and $e^{j2\pi f_k t}$ is the IFFT sinusoid. The term $1/\sqrt{N}$ in front retains power and provides symmetry between the operations. Normally, $1/N$ normalization is used in one direction and unity in the other instead.

The orthogonality of the signal can be guaranteed by

$$f_k = f_0 + k\Delta f \qquad (3.7)$$

Generally, the PAPR of transmitting the OFDM signal is defined as the ratio between the peak power and average power [4]:

$$PAPR = 10\log\left(\frac{P_{\text{peak}}}{P_{\text{avg}}}\right) dB \qquad (3.8)$$

The complex baseband signal as in Eq. (3.6) is defined as over time interval, and, mathematically, PAPR is given as [4]

$$PAPR = 10\log\left(\frac{\max|s(t)|^2}{E|s(t)|^2}\right) dB \qquad (3.9)$$

where $\max|s(t)|^2$ denotes the maximum instantaneous power and $E|s(t)|^2$ denotes the average power of the signal. Thus, the OFDM average signal power can be calculated by [3]

3.1 System Model of OFDM System

$$E = \frac{\text{Sum of the magnitude of all OFDM symbol}}{\text{No. of OFDM symbol}(N)} \quad (3.10)$$

Proof The average signal power in (t) is

$$\begin{aligned}
E &= \frac{1}{\sqrt{N}} \int_0^T |s(t)|^2 dt \\
&= \frac{1}{\sqrt{N}} \int_0^T s(t)\tilde{s}(t) dt \\
&= \frac{1}{\sqrt{N}} \int_0^T \left[\sum_{k=0}^{N-1} X_k \cdot e^{j2\pi f_k t}\right]\left[\sum_{k=0}^{N-1} \tilde{X}_k \cdot e^{-j2\pi f_k t}\right] dt \\
&= \frac{1}{\sqrt{N}} \sum_{k=0}^{N-1}\sum_{k=0}^{-1} X_k \cdot \tilde{X}_k \int_0^T e^{j2\pi f_k t} dt \\
&= \sum_{k=0}^{N-1} |X_k|^2 \quad (3.11)
\end{aligned}$$

From the central limit theorem, the real and imaginary values of (t) become Gaussian distributed with the condition that the N value is large ($N > 64$). Therefore, the amplitude of the OFDM signal has a Rayleigh distribution, with a cumulative distribution given by [5]

$$F(\delta) = 1 - e^{\delta} \quad (3.12)$$

The probability that the PAPR is below a certain threshold level, δ, can be written as [5]

$$P(\text{PAPR} < \delta) = (1 - e^{-\delta})^N \quad (3.13)$$

The performance of a PAPR reduction technique is usually evaluated using the complementary cumulative distribution function (CCDF) by estimating the probability of PAPR of the OFDM symbol that surpasses the threshold level, δ, [6]:

$$\text{CCDF}(\delta) = P(\text{PAPR} > \delta) = 1 - (1 - e^{-\delta})^N \quad (3.14)$$

Equation (3.14) elucidates the relationship between the CCDF of PAPR and the number of subcarriers, N, in the OFDM systems. These two parameters are directly proportional to each other; wherefore, an increase in the number of subcarriers will lead to an increase in CCDF of PAPR.

Planning and optimizing wireless communication systems requires careful consideration of SNR and PAPR parameters. Techniques like channel coding, modulation methods, and power amplifier design aim to increase SNR and decrease PAPR. SNR directly impacts bit error rate (BER) and system performance, while PAPR affects power amplifier efficiency and linearity [7]. Analyzing these metrics ensures reliable and efficient wireless communication.

3.1.2 Bit Error Rate (BER) Analysis Through OFDM Signal Equation

The transmission and reception of data without any error are one of the intentions of any communication system. Therefore, end-to-end performance measurements are needed to prove the reliability of a communication system in terms of how well the output signal reproduces the input. BER is a fundamental measure of digital communication system quality, which measures bit errors in a given number of bit transmissions.

The choice of modulation scheme significantly influences BER, with advanced schemes like QAM and PSK offering higher data throughput and spectral efficiency. Optimizing the modulation scheme to match channel characteristics can improve BER [8]. Understanding different digital modulation techniques can balance data throughput, spectral efficiency, and error performance. Using adaptive modulation and robust error correction methods can enhance data transmission reliability, ensuring efficient and effective communication.

The concept of bit error rate, or BER, may be simplified into the following mathematical relationship:

$$BER = \frac{Number of errors}{Total number of bits sent} \qquad (3.15)$$

By definition, the BER can be written as [9]

$$BER = \frac{Bit_{error}}{Bit_{total}} \qquad (3.16)$$

where Bit_{error} is the number of bits received in error and Bit_{total} is the total number of received bits. In general, BER is expressed as 10 to a negative power. For example, if a transmission possesses a BER of 10^{-4}, this implies that out of 10,000 bits transmitted, 1 had an error. The measurement of BER also can be expressed in percentages as shown in Eq. 3.16:

$$BER = \left(\frac{Number of errors}{Total number of bits sent}\right) \times 100\% \qquad (3.17)$$

For an OFDM system to be reliable and efficient for various applications, its BER performance is a significant factor. Usually, a percentage is used to represent it. By considering the factors that affect BER and employing appropriate techniques, it is possible to achieve high-quality data transmission over wireless channels.

The BER performance against signal-to-noise ratio (SNR) in wireless communication is the well-known performance criterion to plot the BER curve. While reducing the PAPR CCDF is the main objective of PAPR reduction techniques, this is typically attained at the cost of increasing the BER. At some point, the HPA output signal is eventually cut off relative to the average power (clipping level), resulting in substantial in-band distortion. Under the probabilistic techniques category, the transmission of side information (SI) and the OFDM symbol becomes crucial. In the worst-case scenario, the received SI could be corrupted when it arrives at the receiver and consequently jeopardize the entire OFDM symbol; in other words, it degrades the BER performance [10].

Theoretically, the BER performance of an M-QAM modulation technique with Rayleigh fading can be calculated using the following formula [11]:

$$P_{b,MQAM} = \frac{2\left(\sqrt{M}-1\right)}{\sqrt{M}\log_2 M}\left(1 - \sqrt{\frac{\frac{3\log_2 M}{M-1}\frac{E_b}{N_0}}{2 + \frac{3\log_2 M}{M-1}\frac{E_b}{N_0}}}\right) \qquad (3.18)$$

where $\frac{E_b}{N_0}$ is (energy per bit to noise power spectral density ratio) which is a crucial metric in communication systems that quantifies the signal-to-noise ratio (SNR) at the receiver.

In this study, BER performance in Rayleigh fading channel is demonstrated to verify that the BER performance of the proposed technique can be maintained as in conventional OFDM. This proves that the proposed technique does not result in BER degradation when applied to the OFDM system.

References

1. A.M. Jaradat, J.M. Hamamreh, H. Arslan, Modulation options for OFDM-based waveforms: classification, comparison, and future directions. IEEE Access **7**, 17263–17278 (2019). https://doi.org/10.1109/ACCESS.2019.2895958
2. Y.S. Cho, J. Kim, W.Y. Yang, *MIMO-OFDM Wireless Communications with MATLAB* (John Wiley & Sons, 2010)
3. E. Abdullah, A. Idris, A. Saparon, Minimizing high PAPR in OFDM system using circulant shift codeword. J. Teknol. **78**(2), 135–140 (2016) [Online]. Available: www.jurnalteknologi.utm.my
4. M. Razaul, S. Mushfiq, M. Asif, Comparative analysis of various wireless digital modulation techniques with different channel coding schemes under AWGN channel. Int. J. Comput. Appl. **161**(3), 30–34 (2017). https://doi.org/10.5120/ijca2017913139
5. V.V. Panicker, A. Najia, A. Koshy, A distortion based iterative filtering and clipping for better PAPR reduction in OFDM, in *ICCISc 2021—2021 International Conference on Communication, Control and Information Sciences, Proceedings*, vol. 1 (2021), pp. 1–6. https://doi.org/10.1109/ICCISc52257.2021.9485004

6. A.S.B.V.M. Kulkarni, An overview of various techniques to reduce the peak-to-average power ratio in multicarrier transmission systems, in *IEEE International Conference on Computational Intelligence and Computing Research* (2012)
7. K. Savci, G. Galati, G. Pavan, Low-PAPR waveforms with shaped spectrum for enhanced low probability of intercept noise radars. MDPI Remote Sens. **13**(2372), 1–25 (2021). https://doi.org/10.3390/rs13122372
8. K. Singh, A.V. Nirmal, Overview of modulation schemes selection in satellite based communication. ICTACT J. Commun. Technol. **11**(3), 2203–2207 (2020). https://doi.org/10.21917/ijct.2020.0326
9. T. Patra, S. Sil, Bit error rate performance evaluation of different digital modulation and coding techniques with varying channels, in *2017 8th Industrial Automation and Electromechanical Engineering Conference, IEMECON 2017* (2017), pp. 4–10. https://doi.org/10.1109/IEMECON.2017.8079551
10. Y. Rahmatallah, S. Mohan, Peak-to-average power ratio reduction in OFDM systems: a survey and taxonomy. IEEE Commun. Surv. Tutorials **15**(4), 1567–1592 (2013). https://doi.org/10.1109/SURV.2013.021313.00164
11. A. Farzamnia, M. Mariappan, E. Moung, R. Thangasalvam, *BER Performance Evaluation of M-PSK and M-QAM Schemes in AWGN, Rayleigh and Rician Fading Channels*, vol. 371 (Springer, Warsaw, 2022). https://doi.org/10.1007/978-3-030-74540-0_11

Chapter 4
PAPR Reduction Techniques in OFDM System

Over the past few decades, researchers have developed many new approaches to deal with the serious PAPR problem in the OFDM system. Since then, numerous papers and articles have proposed different strategies and addressed several issues related to this problem. Generally, PAPR reduction techniques can be classified into three major categories: signal distortion techniques, coding techniques, and probabilistic or multiple signaling techniques [2]. Each of the methods has its strengths and weaknesses. Moreover, the selection of PAPR reduction techniques varies according to system requirements. It depends on various factors such as system capacity reduction, data rate loss, increment in computational complexity, and worsening BER performance [2].

4.1 Signal Distortion Technique

Distorting the signal before amplification forms the basis of the signal distortion technique such as clipping and companding technique to reduce high peaks in the OFDM signal [2, 3]. Simplicity is known to be the major advantage of this technique. Therefore, signal distortion methods shall not bear additional side information. Still, instead, this technique would cause both in-band and out-of-band interference to be introduced in the system along with complexity [4].

4.1.1 Clipping Technique

The fundamental principle of the signal distortion technique is to minimize high peaks in the transmitted OFDM signal by distorting it before the amplification process takes place in HPA. Clipping technique is one of the methods categorized under signal distortion technique renowned for their simple implementation [5]. The operation of this technique is very straightforward in which the high peak of the OFDM signal is clipped before passing it through the HPA. This idea limits the input signal peak envelope to a pre-determined clipping level; otherwise, the input signal passes without interruption. The output signal can be expressed as [72]

$$d(x) = \begin{cases} d(x) \text{ if } |d(x)| < \vartheta| \\ \vartheta e^{j\psi(x)} \text{ if } |d(x)| \geq \vartheta| \end{cases} \quad (4.1)$$

where $d(x)$ is the OFDM signal, $\psi(x)$ is the phase of x, and P is the clipping level, which can be calculated by [5]

$$\vartheta = \sigma \cdot \sqrt{P_{avg}} \quad (4.2)$$

where P_{avg} denotes the average power while σ denotes the clipping ratio which is related to the desired PAPR. Although the clipping technique guarantees peak reduction, it introduces both in-band distortion and out-of-band radiation, which is the major disadvantage of this technique. In-band distortion causes a degradation in BER performance, while the out-of-band radiation reduces spectral efficiency. On the other hand, filtering the clipped OFDM signal can reduce the effect of out-of-band radiation but not for in-band distortion. Therefore, the clipped OFDM signal needs to be filtered to reduce the impact of out-of-band radiation but not for in-band distortion [6]. Besides, the OFDM signal that undergoes the clipping and filtering process tends to experience peak power re-growth, exceeding the clipping level at some points. For this reason, the utilization of iterative clipping and filtering is needed to achieve the desirable PAPR reduction, but in exchange, the computational complexity of the system is increased [7].

4.1.2 Coding Technique

Suppose a total number of N signals with the same phase are added, then a high peak power is generated, equivalent to N times the average power. Certainly, not all codewords end in a bad PAPR. Therefore, when certain measures are taken to minimize the probability of the same phase of the signals, which is the main concept of the coding schemes, a better PAPR reduction can be achieved [8]. Among all the proposed PAPR reduction methods, the coding technique is the most attractive due to the unique feature of error control capability. Reed-Muller codes, Turbo codes, and low-density parity check (LDPC) codes are several codes that researchers have proposed.

The first attempt to reduce PAPR using the block coding technique was introduced by Jones et al. in 1994, where its basic principle is to map three bits of data into four bits of codeword. Here, the first three bits of the codeword are occupied by the three

4.1 Signal Distortion Technique

data bits, while the fourth bits are for the odd parity check bit [8]. This approach manages to achieve significant PAPR reduction through a comprehensive search. The effectiveness of this approach has motivated other researchers to design a better code. However, the exhaustive search that consumes a large memory space and the error correction problem that the author is not addressing have become the limitation for the first approach [9]. Later, an initiative measure has been taken by Tarokh and Jafarkani [10] in 2000 by proposing an approach that leads to a computationally efficient algorithm. This method implemented a geometrical orientation of codewords with a particular phase shift to the coordinates applied at both the transmitter and receiver. For the first time, this paper highlighted the effects of high PAPR due to computational codewords closely related to minimum distance decoding. Besides, further research conducted by Peterson reveals that a small Nyquist rate would lead to reduced PAPR [11].

Reed-Muller Code with Golay Complementary series to reduce PAPR in [12] has become the famous class of code designed for this purpose. Its achievement drives the popularity of this approach in achieving at most 3 dB Crest Factor (CF) and having efficient encoding and good error-correcting capability [13]. However, this code suffers from major rate loss, and its application is applicable for a small number of carriers [72]. Moreover, in terms of performance, Reed-Muller codes are still far from the Shannon limit demand, especially when compared with capacity-achieving codes such as Turbo codes or LDPC codes.

The motivation and determination to find other methods to reduce PAPR using block coding techniques were continued by Al-Akaidi and Daoud with the proposed Turbo codes [14]. Despite maintaining the PAPR low performance, Turbo codes can also achieve a bit error rate relatively close to the Shannon limit [15]. After several years, the researcher saw the potential of LDPC codes and their capability to solve high PAPR issues in the OFDM system. Gallager proposed LDPC codes in 1962, and it was the only code that can approach Shannon's Limit during that time [16]. Unfortunately, it was abandoned for almost thirty years due to limited computational power at that time until it was reintroduced by Mackay in 1999 [17].

Daoud and Alani were the first people who utilized LDPC codes for PAPR reduction, and their research has shown tremendous improvement in dealing with the PAPR problem compared to Turbo codes [18]. In comparison, Turbo codes have low encoding and high decoding complexity, while LDPC codes have high complexity in encoding and low complexity in decoding [19]. Due to its outstanding BER performance in MIMO-OFDM [20], and Wi-Fi and WiMAX [21] system compared to Turbo codes, LDPC codes will become the suitable choice for Forwarding Error Check (FEC) in the future. Despite all the advantages LDPC codes possess, some trade-off still needs to be solved when implementing them, such as encoding complexity, encoding speed, and diversity order. Although many LDPC constructions and encoding methods have been proposed, most of the objectives are not specifically focusing on PAPR reduction purposes. Daoud et al. worked on irregular LDPC code construction by tackling two main constraints in constructing good LDPC code: an approximate upper triangular form with a small gap matrix, cg, and the block-structured matrix feature. This technique offers lower computational complexity along with a reduction in PAPR when compared with Turbo codes [18].

However, Vimal and Kumar, through their studies, have reported that Quasi Cyclic LDPC (QC-LDPC) codes give promising results in PAPR reduction in OFDM system compared with LDPC and Turbo codes [22]. The construction of QC-LDPC codes is based on the circulant matrices concept that allows parallel encoding and decoding. Still, in return, the encoding complexity and encoding speed have to be sacrificed. This technique can minimize the usage of memory size for storing the parity check matrix, decrease the encoding complexity, and increase the encoding speed [23]. However, the author didn't touch on the error-correcting performance at the receiver to see the system's overall performance. Furthermore, there is still a lack of information regarding the QC-LDPC code construction discussed in detail, whereas it is essential to understand the proper shift value. The encoded data, C, can be expressed as [22]

$$C = S_1 R_1 R_2 \quad (4.3)$$

where S_1 is the original block data while $R_1 R_2$ is the code for encoding. For this application, the code rate is 0.5. Up to now, far too little attention has been paid by researchers to proper shift value in QC-LDPC codes toward the reduction of high PAPR. Besides, there are certain drawbacks associated with this technique such as the unsatisfactory BER performance [24] and application of a lower coding rate [23].

4.1.3 Multiple Signaling and Probabilistic Techniques

Multiple signaling technique is a technique that generates a permutation of the multi-carrier signal, where the signal with the minimum PAPR is chosen for transmission. Meanwhile, probabilistic techniques modify and optimize various parameters in the OFDM signal to minimize the PAPR [2]. The benefits of these techniques are that they do not distort the transmitted signal and achieve a substantial reduction in PAPR. However, they also suffer from certain drawbacks, such as a loss in data rate due to the transmission of several side information bits and increased computational complexity.

Multiple signaling and probabilistic techniques work in one of two ways. The first way specifically focuses on the generation of permutation candidates. Generally, this approach is referred to as the interleaving, scrambling, or permutation technique. Meanwhile, the second way concentrates on the generation of modified OFDM signals through multiplication with optimized phase factor or precoding matrix, commonly known as selective mapping (SLM) and precoding technique, respectively.

In multiple signaling and probabilistic techniques, the transmission of side information (SI) is necessary alongside the selected OFDM signal since the final transmission signal is selected from multiple candidates. The information contained in the

4.1 Signal Distortion Technique

side information indicates which candidate was selected from among all candidates. At the receiver, the data is extracted based on the information carried by the side information index. For data recovery purposes, it is extremely important to maintain the reliability of the side information to avoid misinterpretation of the side information. If side information is misinterpreted due to interference, the entire data will be decoded incorrectly, resulting in BER degradation [25, 26].

Apart from the issue mentioned above, multiple signaling and probabilistic techniques also suffer from high computational complexity. The number of candidates and subcarriers, N, are two factors influencing the ability to minimize PAPR for multiple signals and probabilistic techniques [27]. This fact is supported by simulations showing that a large number of candidates contribute to the improvement of PAPR reduction performance. However, things go the opposite way in SLM, where linear growth of PAPR reduction is achieved when the number of candidates ≤ 8 [28]. As the number of candidates exceeds 8, the growth of PAPR reduction becomes sluggish, whereas the desired PAPR value is not yet achieved. Therefore, the high computational complexity in SLM can be explained by the fact that this technique requires more candidates to obtain desirable PAPR reduction. An increase in candidates' numbers will lead to an equivalent rise in IFFT computations [29].

Nowadays, researchers tend to propose a joint technique among multiple signaling and probabilistic schemes that offer simple implementation and significant PAPR reduction with minimum computation operation. By taking that into account, this research focuses on multiple signaling and probabilistic techniques, particularly interleaving and precoding as a new solution for reducing high PAPR with low computational complexity. At the same time, it prevents BER degradation where permutation becomes its basic working operation. Thus, this new technique offers a new solution to reducing candidates, where a detailed explanation will be discussed in the next section.

4.1.3.1 Interleaving Technique

The OFDM candidates in the interleaving or permutation technique bear the same information as the original signal. In terms of concept, there is not much difference between interleaving and SLM, except the generation of multiple OFDM candidates, K, in the interleaving technique through permutation of data based on certain formulation. Van Eetvelt et al. in [30] were the first to introduce the permutation technique by using four scrambled candidates. The authors reported that this approach could reduce PAPR CCDF by 2% of the maximum possible value. For this reason, the scramble formulation is vital to ensure the maximum probability of a minimum CCDF PAPR value which can be achieved. The permutation process can occur either in bit or symbol form, in which bit permutation is executed before the digital modulation process.

In contrast, symbol permutation is executed after the digital modulation process. Moreover, Jayalath et al., through research in [31, 32], found that symbol permutation tends to degrade the PAPR performance when the K value is greater than

4. Unlike symbol permutation, bit permutation demonstrates stability and consistency to provide a better PAPR. Since more than a decade ago, there have been different kinds of interleaving formulas suggested for combating serious PAPR problems, such as pseudo-random, spread interleaver, convolutional, periodic, odd–even symmetric interleaver [33], and polyphaser interleaving and inversion [32]. Among those formulas, pseudo-random is the superior one with the advantage of better PAPR reduction despite sacrificing its BER performance. Theoretically, the achievable PAPR by interleaving technique is derived in [34].

With the intention to reduce computational complexity, Marsalek [35] proposed a technique that utilizes the adaptive symbol selection principle with several replicas through the use of several interleavers. This formulation work is integrated within the IFFT block where complex multiplication is needed and estimated as [35]

$$N_{mpy,c} = \left(\frac{N}{2}\log_2 N\right)K - \left(\frac{N}{2}\log_2 M\right)(K-1) \qquad (4.4)$$

where M is the length of elementary IFFT blocks. There is no doubt that this technique shows great potential in minimizing the number of candidates to six. However, the complex multiplication involved in the process is presumed to increase the overall system's complexity.

Meanwhile, Lu et al. in [36] have come up with the idea of utilizing a set of cyclically shifted phase sequences (CSPS) in the time domain and required the usage of one IFFT block only. This method, in which the interleaving takes place after the IFFT process, effectively reduces high PAPR. However, the authors never mentioned the BER performance in the study. Later, a new permutation formula based on SLM called the data-position permutation (DPP) technique was suggested by Wen et al. in [37]. In this method, the generation of the OFDM candidates is performed through the permutation and rotation process of the symbol data before continuing to the IFFT modulation process. Overall, the performance of DPP in terms of PAPR reduction is very satisfactory compared to SLM. However, when it comes to BER performance, the result shows severe degradation.

The formulation of data permutation is very complicated and needs to be well-designed. For example, if the data is a position too far from the original sequence, it might cause the BER to be degraded. Instead, if it is too close, no substantial improvement in PAPR can be achieved. Thus, researchers are still searching for the best formulation of permuting the data to balance PAPR reduction and BER performance.

4.1.4 Selective Mapping Technique

Since it was first introduced in 1996 by Bäuml [38], the popularity of SLM techniques as one of the most interesting and effective schemes to reduce PAPR has received considerable attention from researchers. The fundamental concept of SLM

4.1 Signal Distortion Technique

is to generate a set of sufficiently different modified OFDM symbols, all of which carry the same information as the original OFDM symbol and select the one with minimum PAPR for transmission. In a conventional OFDM system utilizing the SLM approach for PAPR reduction, the alternative symbol sequences are generated by multiplying the input data block, $Y = [Y_0, Y_1, \ldots, Y_{N-1}]^T$, with different phase factors, $P^u = [P^1, P^2, \ldots, P^U]^T$, $u = 1, 2, \ldots, U$, as shown in Fig. 4.1. For the u th phase factor, the alternative symbol sequences can be denoted as $Y_u = [Y_1 P^1, Y_2 P^2, \ldots, Y_{N-1} P^U]^T$. Therefore, the alternative multicarrier OFDM signal can be written as [39]

$$y^u = \frac{1}{\sqrt{N}} \sum_{k=0}^{N-1} Y^u \cdot e^{j2\pi f_k t} \quad 0 < t < NT \tag{4.5}$$

Finally, the PAPR of the y^u signals are evaluated separately and the one with minimum PAPR is selected for final transmission.

For the receiver to recover the original data sequence, the conventional SLM technique needs to transmit the index u that identifies the selected phase sequence P^u as side information. Since U number of IFFT blocks is needed in the implementation of the SLM technique, therefore, this technique requires $\log_2 U$ bits of side information embedded into every single alternative symbol's sequence. As addressed in Sect. 4.1.3, there are two limitations associated with the SLM technique: the requirement for transmitting the information about the selected candidate to the receiver as side information (SI) and high computational complexity due to the use of a high number of IFFT blocks. In recent years, the research trend has paid close attention to overcome the drawbacks of the conventional SLM technique. To that end,

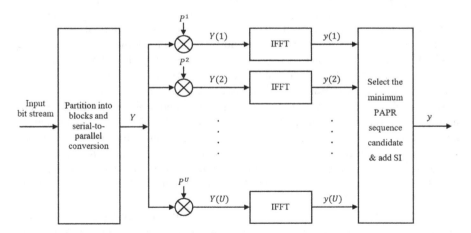

Fig. 4.1 Block diagram of conventional SLM technique (adapted from [40])

researchers have studied the implementation of the SLM technique without the transmission of side information, also known as blind SLM and SLM techniques with low computational complexity.

To avoid the transmission of side information, the author of [41] studied utilizing the maximum likelihood (ML) decoder for the SLM technique. The result shows that the BER performance is equivalent to the conventional SLM, assuming perfect side information recovery. However, at the receiver, it induces substantial decoding complexity. On top of that, the author fails to provide information regarding the PAPR performance. Meanwhile, Joo et al. in [42] proposed a new blind SLM scheme that exploits block partitioning and phase rotation as a method to embed side information into the alternative symbol sequence. In this method, the utilization of a low computational complexity ML decoder is derived to ensure the detection failure probability of side information is kept to a minimum. This ensures achieving the same BER performance compared with the conventional blind SLM scheme in [41]. In terms of PAPR reduction performance, the authors claim it has identical performance as the conventional SLM. A few years later, Joo et al. proposed a new blind SLM scheme in [42]. This time around, the side information is embedded into each phase sequence by giving the phase offset to the elements of the phase sequence. The biorthogonal vectors pre-determine the phase offset for the partitioned sub-blocks. In the pursuit of reducing the decoding complexity, this method makes ML decoder with low decoding complexity as its solution. The numerical results indicate that this approach almost has the same BER and PAPR reduction performance as the conventional blind SLM in [41]. However, in each sub-block, this approach demanded twice the multiplication of the phase factor to the OFDM symbol.

A new data recovery process without side information with low decoding complexity using a pilot phase sequence was proposed by Park et al. in [43]. This proposed technique possesses approximately the same BER performance as the conventional SLM and is considerably better than the ML decoder. For the PAPR reduction performance, it is on par with the conventional SLM. In [44], the authors suggested using a phase rotation sequence estimation based on the minimum Euclidean distance of the fourth-power constellation for blind SLM. By implementing the fourth-power constellation, this proposed technique can reduce the calculation of minimum Euclidean distance for symbol candidates and thus leads to lower complexity. To further reduce the complexity and at the same time maintain high estimation accuracy, the authors introduced a set of phase rotation sequences generated by random selection from $\{0°, 135°\}$. The simulation result demonstrates that the proposed idea can significantly reduce the complexity and preventing BER degradation. Unfortunately, this method was experimented with using single-carrier signals, and therefore, the true potential of this method in OFDM signals remains unknown.

Another blind SLM technique with low complexity characteristic was proposed in [45]. A well-designed phase rotation vector in the time domain at the transmitter is the main idea for this method which can be regarded as an equivalent wireless channel without side information transmission. The effect of phase rotation vectors can be removed at the receiver by the conventional channel estimation method. The authors

4.1 Signal Distortion Technique

reported that the proposed scheme could achieve better PAPR reduction performance with a slight increase in BER performance. Helmert-SLM is the technique proposed by Wei et al. in [46] to reduce PAPR in conventional SLM, whose phase vector is chosen from the row of the Helmert matrix. Due to the Helmert matrix's fixed structure, the transmission of side information becomes unnecessary for this technique. Simulation results indicated that Helmert-SLM achieved remarkable improvement in the reduction of PAPR compared with other SLM techniques that used Hadamard [47] and Riemann matrix [48]. However, the author's research did not cover the BER performance and, also, there is no improvement on computational complexity reported in the study.

Hidden side information seems to be another approach in the blind SLM scheme through a method of combining SI with pilot tones and embedding the SI in the transmitted data [49–52]. Unfortunately, the phase rotation information that is too tightly coupled with the SI and limited PAPR reduction performance become the challenges of these techniques. To tackle these issues, a new method that separates the phase rotation information from side information and uses the Hadamard codes as the phase rotation information is proposed in [53]. Even though the proposed Hadamard-based SLM (HSLM) technique can achieve a significant PAPR reduction and improvement in BER performance, it is only applicable to binary phase shift keying (BPSK). The research effort is then continued by Li et al. in [54] through a proposed of semi-Hadamard matrix generation method as a new side information hiding technique. The authors compared the results with the conventional HSLM. Their proposed method outperforms the conventional HSLM in terms of PAPR performance, and this method can be used for both BPSK and QAM with a good BER performance. On the other hand, computational complexity is another trade-off of the SLM technique that needs to be worried about. This problem is associated with an increase in the number of candidates, U, for improving the PAPR reduction capability of SLM, which indirectly leads to an increase in IFFT computations. The major concern in conventional SLM is to reduce the high PAPR in OFDM systems by increasing the number of candidates without having any proportional rise in the number of IFFT blocks. Therefore, Fischer, through his research in [55], has designed an improved SLM scheme that employs the principle of widely linear signal processing that treats the real and imaginary parts of the data separately to generate U^2 number of candidates. This method shows better PAPR performance than conventional SLM employing the same complexity but nothing is being mentioned regarding the BER performance of this proposed technique. In [56], a new SLM scheme called GreenOFDM is presented, capable of generating more candidates while maintaining the same number of IFFT computations as in conventional SLM. In this scheme, the input data is divided equally into two groups, each containing $U/2$ candidates and finally added to generate $(U^2/4)$ set of candidates. It is reported that this scheme outperforms conventional SLM, and, once again, the evaluation of BER performance is neglected. A few years later, a variant of the GreenOFDM algorithm is proposed in [29] where the number of generated candidates can be increased to $(U + U^2/4)$. Both simulation and analytical results demonstrated that the proposed method surpasses the PAPR performance of conventional SLM and GreenOFDM with a minimal downgrade effect on BER performance.

For the sake of achieving a substantial reduction in computational complexity, a modified SLM (M-SLM) technique has been proposed by Hu et al. in [57]. Such a reduction can be achieved by using two sets of orthogonal phase factors to generate two groups of original OFDM candidate signals. Then, based on the linear property of IFFT, more candidate signals are obtained by employing the linear combinations of candidate signals from the above two groups. Although the computational complexity is successfully reduced, it is a huge loss when no reduction in PAPR is achieved as the M-SLM owned similar PAPR performance with original SLM and GreenOFDM. High computational complexity in SLM can also be seen from a different perspective. A full search for finding the smallest PAPR among candidate signals could also contribute to this problem. Hence, through his initiative, Joo [1] has proposed an early termination algorithm to immediately terminate the searching process when a candidate signal whose PAPR is less than a pre-determined target PAPR threshold value is identified. The target PAPR threshold is set not to increase the clipping probability of the transmit power amplifier. The utilization of this technique allowing the trade-off between PAPR performance and computational complexity can be controlled by adjusting the threshold.

The researcher may have learned from experience or literature review that the good PAPR performance of the SLM technique comes at a price of high computational complexity due to the high number of IFFT blocks associated with the number of candidates generated. Based on the studies discussed so far and the consideration given to all evidence, it seems that there is a need to obtain a substantial PAPR reduction with a minimum number of candidates used along with high-reliability side information. Therefore, this research addresses high computational complexity by proposing a new PAPR reduction technique with a minimum number of candidates generated, thus minimizing the IFFT block's usage.

4.1.5 Selective Codeword Shift Technique

Selective codeword shift (SCS) proposed in [1] is newly categorized under the interleaving technique. A simple but effective permutation formulation proposed by the author has proven to impact PAPR reduction performance significantly. Circulant codeword shift formulation acts as the fundamental for this proposed technique that allowed it to achieve good PAPR reduction while at the same time maintaining the same BER performance as conventional OFDM. Besides, the computational complexity of this technique is quite low as it required the minimum use of IFFT block. Generally, the basic idea of this technique is to generate a set of sufficient different modified symbol candidates, $X(s)$ where $0 \leq s \leq m-1$ and m is the number of bits per symbol. Each candidate possesses the same information as the original OFDM symbol, X, and the one with minimum PAPR is selected for transmission.

4.1 Signal Distortion Technique

Unlike the SLM technique in which the alternative symbol sequences are generated through the multiplication of phase factor, in the SCS technique, as seen in Fig. 4.2, the alternative symbol sequences are generated through the shifting process performed to the codeword bits. The total number of generated alternative codeword sequences is $m - 1$. This shifting process takes place between serial-to-parallel conversion and QAM modulation. The original symbol sequences, $X = [X(0), X(1), \ldots, X(N-1)]^T$, are shifted according to the desired shift factor, S^ζ where $0 < \zeta < m-1$. The total number of shift factors is equivalent to the number of alternative codeword sequences. Thus, the alternative codeword sequences are given as [1]

$$C^S = \prod_{k-1}^{K} C_k \otimes S^\zeta 0 \leq \zeta \leq m - 1 \quad (4.6)$$

where C is the codeword and S is the number of alternative codewords produced by using the shift factor, S^ζ. For instance, the codeword bit position for a 64 QAM with 6 bits per symbol after undergoing the SCS process is shown in Table 4.1. The table shows the creation of an alternative codeword from the original codeword sequence for a different number of shift factors. Then, the binary data are modulated using a QAM modulator to produce OFDM symbols in complex numbers as a preparation to be modulated by IFFT. The output of symbol sequences is $Z_S = [Z_{1,\zeta}, Z_{2,\zeta}, \ldots, Z_{K,\zeta}]$, where K is the number of sub-blocks.

The process is followed by IFFT modulation to produce S number of OFDM signals in the time domain, and the signal can be represented as [1]

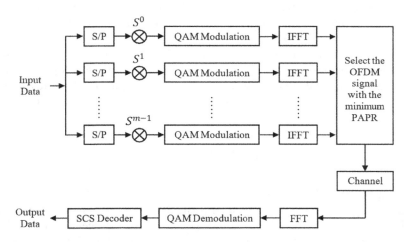

Fig. 4.2 Block diagram of SCS technique in OFDM system (adapted from [1])

Table 4.1 SCS bit arrangement for different number of shift factors using 64-QAM [1]

Codeword	Shift factor, S^ζ	Bit position
Original	–	$[c_1, c_2, c_3, c_4, c_5, c_6]$
Alternative Codeword 1	1	$[c_6, c_1, c_2, c_3, c_4, c_5]$
Alternative Codeword 2	2	$[c_5, c_6, c_1, c_2, c_3, c_4]$
Alternative Codeword 3	3	$[c_4, c_5, c_6, c_1, c_2, c_3]$
Alternative Codeword 4	4	$[c_3, c_4, c_5, c_6, c_1, c_2]$
Alternative Codeword 5	5	$[c_2, c_3, c_4, c_5, c_6, c_1]$

$$z^S(t) = \frac{1}{\sqrt{N}} \sum_{k=0}^{N-1} Z^S_k \cdot e^{j2\pi f_k t} \quad (4.7)$$

Lastly, the minimum PAPR among the S number of OFDM signals is selected for transmission. The transmitted OFDM signal is obtained by [1]

$$S^S(t) = \mathrm{argminPAPR}(z^S(t)) \quad (4.8)$$

It must be mentioned here that side information needs to be included with the transmitted signals to retrieve the data at the receiver properly.

The author also demonstrates that SCS is applicable in MIMO-OFDM systems as an effective solution for reducing serious PAPR problems through the study in [58, 59]. Furthermore, the author also reported a remarkable improvement in BER performance when SCS is tested in the MIMO-OFDM system. However, overall, the PAPR reduction capability of the SCS technique can still be improved as other permutation formulations are still not being explored. Here, codeword properties can be manipulated to obtain a better PAPR reduction. Furthermore, the relationship between codeword structure and PAPR reduction capability is still not studied. Thus, it is not obvious which codeword structure will give good PAPR performance, whether altered or unaltered codeword structure. Besides, the need for transmitting side information to the receiver has caused SCS to experience a slight data rate loss.

4.1.6 Precoding Technique

Precoding is one of the methodologies that have been proposed for the reduction of PAPR in the literature. As a constellation shaping method that involves the unitary matrix transformations [60, 61], the key idea of the precoding scheme is to transform the modulated data through the multiplication with an $N \times N$ precoding matrix before the operation of IFFT. Since the precoding matrix, P, is a unitary matrix, therefore, it satisfies the following relationship [60]:

4.1 Signal Distortion Technique

$$P^*P = 1 \tag{4.9}$$

where I is an identity matrix with $N \times N$ dimensions and P^* is the Hermitian transpose of P. In general, the precoding matrix, P, can be written as [60]

$$P = \begin{bmatrix} P_{0,0} & P_{0,1} & \cdots & P_{0,N-1} \\ P_{1,0} & P_{1,1} & \cdots & P_{1,N-1} \\ \vdots & \vdots & \ddots & \vdots \\ P_{N-1,0} & P_{N-1,1} & \cdots & P_{N-1,N-1} \end{bmatrix} \tag{4.10}$$

where N is the total number of subcarriers and $p_{n,m}$ represents the entry of the nth row and mth column of P. Several different transforms have been used as precoding matrices for PAPR reduction, such as Discrete Fourier Transform (DFT), Walsh-Hadamard Transform (WHT), Discrete Cosine Transform (DCT), and Discrete Hartley Transform (DHT). The entries of several precoding matrices are given in Table 4.2. The precoding technique is an attractive approach due to its simplicity for implementation, and good PAPR reduction performance without experiencing any BER degradation, where no transmission of side information is required to the receiver [61, 62]. However, although this method alleviates high PAPR, its capability to reduce PAPR is limited due to the fixed precoding matrix [63]. Therefore, several methods have been studied by combining the precoding technique with other PAPR reduction techniques to improve its PAPR reduction capability further.

In [65], a joint technique between root-based μ-law companding (RMC) and DHT is proposed to reduce high PAPR in the OFDM system. RMC scheme is known to simultaneously compressing/expanding high/low amplitude OFDM signals, respectively compared with the standard μ-law companding (MC) technique which expands the amplitudes of low power signals without impacting the higher amplitude signals. As a result, this limits the PAPR reduction performance of the MC scheme. The simulation results confirm that the proposed DHT-precoded RMC companding transform achieves better PAPR reduction performance while preserving its BER performance. Regrettably, the complexity analysis is not provided in this study.

Table 4.2 Entries of different precoding matrices [64]

Transform	Precoding matrix entries
DFT	$p_{n,m} = \exp\left(j\frac{2\pi nm}{N}\right)$
WHT	$H_1 = [1]$, $H_2 = \frac{1}{\sqrt{2}}\begin{bmatrix} 1 & 1 \\ 1 & -1 \end{bmatrix}$, $H_{2N} = \frac{1}{\sqrt{2N}}\begin{bmatrix} H_N & H_N \\ H_N & H_N^{-1} \end{bmatrix}$
DHT	$p_{n,m} = \cos\left(\frac{2\pi nm}{N}\right) + \sin\left(\frac{2\pi nm}{N}\right)$

Meanwhile, another effective PAPR reduction technique involving a companding scheme is proposed in [66] by combining piecewise linear companding with DHT transform. In this technique, the encoding process of DHT takes place before the IFFT modulation, while the companding transform is applied after the IFFT modulation. However, even though this proposed scheme has a good PAPR reduction performance, the same thing is not expected for the BER performance and complexity. As a result, the proposed scheme suffers from a slight BER degradation with a small increase in complexity.

Several precoding schemes have been proposed by Sravanti et al. in [67] to be merged with partial transmit sequences (PTS) to identify which joint technique has the finest PAPR reduction. The authors have selected DFT, DHT, and WHT as potential candidates for the precoding technique. However, simulation results show that the DHT-PTS techniques have superiority over others regarding PAPR reduction performance. Therefore, Sravanti et al. in [68] investigate which precoding scheme is better coupled with the SLM technique for PAPR reduction in OFDM systems. The same precoding candidates used in [67] are used in this research, and, once again, DHT has proven to be the best candidate for a joint technique with SLM compared with other precoding schemes. However, since the authors focus only on the PAPR performance in [67, 68], the BER performance of DHT-PTS and DHT-SLM remains a question.

Another improvement in PAPR reduction of DCT was presented in [69]. The enhancement of DCT is done with the help of the interleaving and PTS technique. In this study, the authors proposed two ways on how the DCT encoding process would be performed. For the first way, the encoding process is performed before the IFFT operation, while, for the second way, it is performed after the IFFT operation. It is quite unusual for an encoding process in precoding technique to be performed after the IFFT operation as the encoding process is normally done before the IFFT operation. However, the idea to execute DCT encoding after IFFT modulation can still compete with the PAPR performance of DCT encoding executed before IFFT modulation.

Nowadays, hybrid techniques, which are a combination of several methods, are considered a good option for high PAPR reduction as they produce a new technique with better PAPR reduction capability [70, 71]. But unfortunately, the hybrid scheme also inherits the weaknesses of the individual technique. Therefore, a good strategy is needed for reducing PAPR, which involves the use of a hybrid technique by considering certain factors such as PAPR reduction capability, power increment in the transmitted signal, BER degradation at the receiver, data rate loss, computational complexity, and bandwidth expansion when choosing a specific PAPR reduction method [2].

References

1. E. Abdullah, A. Idris, A. Saparon, Minimizing high PAPR in OFDM system using circulant shift codeword. J. Teknol. **78**(2), 135–140 (2016). [Online]. Available: www.jurnalteknologi.utm.my
2. F. Sandoval, G. Poitau, F. Gagnon, Hybrid peak-to-average power ratio reduction techniques: review and performance comparison. IEEE Access **5**, 27145–27161 (2017). https://doi.org/10.1109/ACCESS.2017.2775859
3. D. Teja Sai Vishnu Vardhan, A. Narendra Kumar, P. Vijaya Kumar, K. Chandra Kiran, G. Jyothiraditya, A. Ravi Raja, Fusion of adaptive SLM technique with companding for PAPR reduction in 5G MIMO-OFDM system, in *2023 2nd International Conference on Electrical, Electronics, Information and Communication Technologies, ICEEICT 2023* (2023), pp. 1–5. https://doi.org/10.1109/ICEEICT56924.2023.10157778
4. Y.P. Tu, Z.T. Zhan, Y.F. Huang, A novel alternating μ-law companding algorithm for PAPR reduction in OFDM systems. MDPI Electron. (Switzerland) **13**(694), 1–22 (2024). https://doi.org/10.3390/electronics13040694
5. B. Tang, K. Qin, X. Zhang, C. Chen, A clipping-noise compression method to reduce PAPR of OFDM signals. IEEE Commun. Lett. **23**(8), 1389–1392 (2019). https://doi.org/10.1109/lcomm.2019.2916052
6. N.M.A.E.D. Wirastuti, N. Pramaita, I.M.A. Suyadnya, D.C. Khrisne, Evaluation of clipping and filtering-based PAPR reduction in OFDM system. J. Electr. Electron. Inform. **1**(2), 18 (2017). https://doi.org/10.24843/jeei.2017.v01.i02.p05
7. J. Armstrong, Peak-to-average power reduction for OFDM by repeated clipping and frequency domain filtering. Electron. Lett. **38**(5), 246–247 (2002). https://doi.org/10.1049/el:20020175
8. F. Sandoval, G. Poitau, F. Gagnon, On optimizing the PAPR of OFDM signals with coding, companding, and MIMO. IEEE Access **7**, 24132–24139 (2019). https://doi.org/10.1109/ACCESS.2019.2899965
9. M. Virdi, Overview of various PAPR techniques in OFDM systems. J. Netw. Commun. Emerg. Technol. (JNCET) **2**(2), 45–48 (2015)
10. V. Tarokh, H. Jafarkhani, On the Computation and reduction of the peak-to-average power ratio in multicarrier communications. IEEE Trans. Commun. **48**(1), 37–44 (2000). https://doi.org/10.1109/26.818871
11. K.G. Paterson, V. Tarokh, On the existence and construction of good codes with low peak-to-average power ratios (2000)
12. J.A. Davis, J. Jedwab, Peak-to-mean power control in OFDM, Golay complementary sequences and Reed-Muller codes, in *IEEE International Symposium on Information Theory—Proceedings*, vol. 45, no. 7 (1998), p. 190. https://doi.org/10.1109/ISIT.1998.708788
13. K.G. Paterson, Generalised Reed-Muller codes and power control in OFDM modulation, in *IEEE International Symposium on Information Theory—Proceedings* (1998), p. 194. https://doi.org/10.1109/ISIT.1998.708792
14. A.A. Abouda, PAPR reduction of OFDM signal using turbo coding and selective mapping, in *Proceedings of the 6th Nordic Signal Processing Symposium—NORSIG 2004*, Espoo, Finland (2004)
15. M. Sabbaghian, Y. Kwak, B. Smida, V. Tarokh, Near Shannon limit and low peak to average power ratio turbo block coded OFDM. IEEE Trans. Commun. **59**(8), 2042–2045 (2011). https://doi.org/10.1109/TCOMM.2011.080111.090356
16. R.G. Gallager, Low-density parity-check codes. IRE Trans. Inform. Theory **8**(1), 21–28 (1962)
17. D.J.C. MacKay, Good error-correcting codes based on very sparse matrices. IEEE Trans. Inf. Theory **45**(2), 399–431 (1999). https://doi.org/10.1109/18.748992
18. O. Daoud, O. Alani, Reducing the PAPR by utilisation of the LDPC code. IET Commun. **3**(4), 520–529 (2009). https://doi.org/10.1049/iet-com.2008.0344
19. T. Velmurugan, S. Balaji, A.S. Rennie, D. Sumathi, Efficiency of the LDPC codes in the reduction of PAPR in comparison to turbo codes and concatenated turbo-Reed Solomon codes

in a MIMO-OFDM system. Int. J. Comput. Electr. Eng. (2013). https://doi.org/10.7763/ijcee.2010.v2.267
20. O. Daoud, Use of LDPC to improve the MIMO-OFDM systems performance, in *2008 5th International Multi-Conference on Systems, Signals and Devices, SSD'08* (2008), pp. 1–5. https://doi.org/10.1109/SSD.2008.4632778
21. S.H. Gupta, B. Virmani, LDPC for Wi-Fi and WiMAX technologies, in *2009 International Conference on Emerging Trends in Electronic and Photonic Devices and Systems, ELECTRO '09* (2009), pp. 262–265. https://doi.org/10.1109/electro.2009.5441120
22. S.P. Vimal, K.R.S. Kumar, PAPR reduction in OFDM systems using quasi cyclic LDPC codes. Asian J. Sci. Res. **6**(4), 715–725 (2013)
23. S. Borwankar, D. Shah, Low density parity check code (LDPC Codes) overview. Inform. Theory (2020). https://doi.org/10.48550/arXiv.2009.08645
24. X. Yang, H.L. Moon, Construction of good Quasi-cyclic LDPC codes, in *IET Conference Publications*, no. 525 (2006), p. 172. https://doi.org/10.1049/cp:20061337
25. S. Eom, H. Nam, Y.-C. Ko, Low-complexity PAPR reduction scheme without side information for OFDM systems. IEEE Trans. Signal Process. **60**(7), 3657–3669 (2012)
26. J. Ji, G. Ren, A new modified SLM scheme for wireless OFDM systems without side information, no. 61072102
27. C.L. Wang, Y. Ouyang, Low-complexity selected mapping schemes for peak-to-average power ratio reduction in OFDM systems. IEEE Trans. Signal Process. **53**(12), 4652–4660 (2005). https://doi.org/10.1109/TSP.2005.859327
28. R.W. Bauml, R.F.H. Fischer, J.B. Huber, Reducing the peak-to-average power ratio of multicarrier modulation by selected mapping. IET Electron. Lett. **32**(22), 2056–2057 (1996)
29. S. Gupta, A. Goel, Improved selected mapping technique for reduction of PAPR in OFDM systems. Int. J. Adv. Comput. Sci. Appl. **11**(10), 117–122 (2020). https://doi.org/10.14569/IJACSA.2020.0111016
30. P. Van Eetvelt, G. Wade, M. Tomlinson, Peak to average power reduction for OFDM schemes by selective scrambling. Electron. Lett. **32**(21), 1963–1964 (1996). https://doi.org/10.1049/el:19961322
31. A.D.S. Jayalath, C. Tellambura, Use of data permutation to reduce the peak-to-average power ratio of an OFDM signal. Wirel. Commun. Mob. Comput. **2**(2), 187–203 (2002). https://doi.org/10.1002/wcm.47
32. Y.S. Su, Computationally efficient PAPR reduction Of SFBC-OFDM signals by polyphase interleaving and inversion, in *International Symposium on Wireless Personal Multimedia Communications, WPMC*, vol. 2015-Janua, no. 1 (2015), pp. 729–733. https://doi.org/10.1109/WPMC.2014.7014911
33. A.D.S. Jayalath, C. Tellambura, Use of interleaving to reduce the peak-to-average power ratio of an OFDM signal, in *Conference Record/IEEE Global Telecommunications Conference*, vol. 1 (2000), pp. 82–86. https://doi.org/10.1109/glocom.2000.891696
34. A.D.S. Jayalath, C.R.N. Athaudage, On the PAR reduction of OFDM signals using multiple signal representation. IEEE Commun. Lett. **8**(7), 425–427 (2004). https://doi.org/10.1109/LCOMM.2004.832767
35. R. Maršálek, On the reduced complexity interleaving method for OFDM PAPR reduction. Radioengineering **15**(3), 49–53 (2006)
36. G. Lu, P. Wu, D. Aronsson, Peak-to-average power ratio reduction in OFDM using cyclically shifted phase sequences. IET Commun. **1**(6), 1146–1151 (2007). https://doi.org/10.1049/iet-com:20060038
37. J.H. Wen, S.H. Lee, C.C. Kung, SLM-based data position permutation method for PAPR reduction in OFDM systems. Wirel. Commun. Mob. Comput. **13**(11), 985–997 (2013). https://doi.org/10.1002/wcm.709
38. R.W. Bäuml, R.F.H. Fischer, J.B. Huber, Reducing the peak-to-average power ratio of multicarrier modulation by selected mapping. Electron. Lett. **32**(22), 2056–2057 (1996). https://doi.org/10.1049/el:19961384

39. E. Abdullah, A. Idris, A. Saparon, Modified selective mapping scheme with low complexity for minimizing high peak-average power ratio in orthogonal frequency division multiplexing system. AIP Conf. Proc. **1774**, 2016 (2016). https://doi.org/10.1063/1.4965092
40. A.A. Barakabitze, Md.A. Ali, Behavior and techniques for improving performance of OFDM systems for wireless communications. Int. J. Adv. Res. Comput. Commun. Eng. **4**(1), 237–245 (2015). https://doi.org/10.17148/ijarcce.2015.4152
41. A.D.S. Jayalath, C. Tellambura, SLM and PTS peak-power reduction of OFDM signals without side information. IEEE Trans. Wirel. Commun. **4**(5), 2006–2013 (2005). https://doi.org/10.1109/TWC.2005.853916
42. H.S. Joo, S.J. Heo, H.B. Jeon, J.S. No, D.J. Shin, A new blind SLM scheme with low decoding complexity for OFDM systems. IEEE Trans. Broadcast. **58**(4), 669–676 (2012). https://doi.org/10.1109/TBC.2012.2216472
43. J. Park, E. Hong, D. Har, Low complexity data decoding for SLM-based OFDM systems without side information. IEEE Commun. Lett. **15**(6), 2011–2013 (2011)
44. A. Boonkajay, F. Adachi, A low-complexity phase rotation estimation using fourth-power constellation for blind SLM, in *IEEE Vehicular Technology Conference*, vol. 2018-August (2018), pp. 1–5. https://doi.org/10.1109/VTCFall.2018.8690569
45. Y. Xia, J. Ji, Low-complexity blind selected mapping scheme for peak-to-average power ratio reduction in orthogonal frequency-division multiplexing systems. MDPI Inform. (Switzerland) **9**(220), 1–11 (2018). https://doi.org/10.3390/info9090220
46. S. Wei, H. Li, G. Han, W. Zhang, X. Luo, PAPR reduction of SLM-OFDM using Helmert sequence without side information, in *Proceedings of the 14th IEEE Conference on Industrial Electronics and Applications, ICIEA 2019* (2019), pp. 533–536. https://doi.org/10.1109/ICIEA.2019.8833926
47. D.W. Lim, S.J. Heo, J.S. No, H. Chung, On the phase sequence set of SLM OFDM scheme for a crest factor reduction. IEEE Trans. Signal Process. **54**(5), 1931–1935 (2006). https://doi.org/10.1109/TSP.2006.871979
48. N.V. Irukulapati, V.K. Chakka, A. Jain, SLM based PAPR reduction of OFDM signal using new phase sequence. Electron. Lett. **45**(24), 1231–1232 (2009). https://doi.org/10.1049/el.2009.1902
49. S.Y. Le Goff, S.S. Al-Samahi, B.K. Khoo, C.C. Tsimenidis, B.S. Sharif, Selected mapping without side information for PAPR reduction in OFDM. IEEE Trans. Wirel. Commun. **8**(7), 3320–3325 (2009). https://doi.org/10.1109/TWC.2009.070463
50. E.F. Badran, A.M. El-Helw, A novel semi-blind selected mapping Technique for PAPR reduction in OFDM. IEEE Signal Process. Lett. **18**(9), 493–496 (2011). https://doi.org/10.1109/LSP.2011.2160720
51. S. Meymanatabadi, J.M. Niya, B.M. Tazehkand, Multiple recursive generator-based method for peak-to-average power ratio reduction in selected mapping without side information. China Commun. **10**(8), 68–76 (2013). https://doi.org/10.1109/CC.2013.6633746
52. A.M. Elhelw, E.F. Badran, Semi-blind error resilient SLM for PAPR reduction in OFDM using spread spectrum codes. PLoS One **10**(5) (2015). https://doi.org/10.1371/journal.pone.0127639
53. A.S. Namitha, S.M. Sameer, A bandwidth efficient selective mapping technique for the PAPR reduction in spatial multiplexing MIMO-OFDM wireless communication system. Phys. Commun. **25**, 128–138 (2017). https://doi.org/10.1016/j.phycom.2017.09.009
54. N. Li, M. Li, Z. Deng, A modified Hadamard based SLM without side information for PAPR reduction in OFDM systems. China Commun. **16**(12), 124–131 (2019). https://doi.org/10.23919/JCC.2019.12.009
55. R.F.H. Fischer, Widely-linear selected mapping for peak- to-average power ratio reduction in OFDM. Electron. Lett. **43**(14), 40–41 (2007)
56. D.J.G. Mestdagh, J.L.G. Monsalve, J.M. Brossier, GreenOFDM: a new selected mapping method for OFDM PAPR reduction. Electron. Lett. **54**(7), 449–450 (2018). https://doi.org/10.1049/el.2017.4743
57. C. Hu, L. Wang, Z. Zhou, A modified SLM scheme for PAPR reduction in OFDM systems, in *ICEIEC 2020—Proceedings of 2020 IEEE 10th International Conference on Electronics*

58. E. Abdullah, A. Idris, A. Saparon, PAPR reduction using SCS_SLM technique in STFBC MIMO-OFDM. ARPN J. Eng. Appl. Sci. **12**(10), 3218–3221 (2017). *Information and Emergency Communication*, no. 3 (2020), pp. 61–64. https://doi.org/10.1109/ICEIEC49280.2020.9152350
59. E. Abdullah, N.M. Hidayat, SCS-SLM PAPR reduction technique in STBC MIMO-OFDM systems, in *IEEE International Conference on Control System, Computing and Engineering (ICCSCE)* (2017), pp. 104–109. https://doi.org/10.1109/ICCSCE.2017.8284388
60. C.Y. Hsu, H.C. Liao, Generalised precoding method for PAPR reduction with low complexity in OFDM systems. IET Commun. **12**(7), 796–808 (2018). https://doi.org/10.1049/iet-com.2017.0824
61. X. Zhu, G. Zhu, T. Jiang, Reducing the peak-to-average power ratio using unitary matrix transformation. IET Commun. **3**(2), 161–171 (2009). https://doi.org/10.1049/iet-com:20080194
62. A. Ali Sharifi, Discrete Hartley matrix transform precoding-based OFDM system to reduce the high PAPR. ICT Express **5**(2), 100–103 (2019). https://doi.org/10.1016/j.icte.2018.07.001
63. A.S. Namitha, P. Sudheesh, Improved precoding method for PAPR reduction in OFDM with bounded distortion. Int. J. Comput. Appl. **2**(7), 7–12 (2010). https://doi.org/10.5120/682-960
64. M. Mounir, I.F. Tarrad, M.I. Youssef, Performance evaluation of different precoding matrices for PAPR reduction in OFDM systems. Internet Technol. Lett. **1**(6) (2018). https://doi.org/10.1002/itl2.70
65. K. Anoh, B. Adebisi, K.M. Rabie, C. Tanriover, Root-based nonlinear companding technique for reducing PAPR of precoded OFDM signals. IEEE Access **6**, 4618–4629 (2017). https://doi.org/10.1109/ACCESS.2017.2779448
66. A. Thammana, M.K. Kasi, Improvement measures of DHT precoded OFDM over WIMAX channels with piecewise linear companding, in *2016 IEEE Annual India Conference, INDICON 2016*, no. 3 (2017), pp. 1–6. https://doi.org/10.1109/INDICON.2016.7839139
67. T. Sravanti, N. Vasantha, Precoding PTS scheme for PAPR reduction in OFDM system, in *Proceedings of IEEE International Conference on Innovations in Electrical, Electronics, Instrumentation and Media Technology, ICIEEIMT 2017*, vol. 2017-Janua, no. 978 (2017), pp. 250–254. https://doi.org/10.1109/ICIEEIMT.2017.8116844
68. T. Sravanti, N. Vasantha, Performance analysis of precoded PTS and SLM scheme for PAPR reduction in OFDM system, in *Proceedings of IEEE International Conference on Innovations in Electrical, Electronics, Instrumentation and Media Technology, ICIEEIMT 2017*, vol. 2017-Janua, no. 978 (2017), pp. 255–260. https://doi.org/10.1109/ICIEEIMT.2017.8116845
69. B.M. Hardas, S.B. Pokle, Analysis of OFDM system using DCT-PTS-SLM based approach for multimedia applications. Cluster Comput. **22**, 4561–4569 (2019). https://doi.org/10.1007/s10586-018-2140-0
70. S.F. Sultana, G. Dharani, U. Anusha, B. Keerthana, Enhancing PAPR reduction in MIMO-OFDM Systems through grey wolf optimization and hybrid algorithms, in *2024 5th International Conference for Emerging Technology, INCET 2024* (2024), pp. 1–8. https://doi.org/10.1109/INCET61516.2024.10593461
71. A.O. Solomon, F.K. Boakye, A.O.S.A. Folasole, E.O. Isaiah, F.O. Malm, Concept of hybrid transform for reduction of peak to average power ratio in orthogonal frequency division multiplexing system. J. Multidiscip. Eng. Sci. Technol. (JMEST) **10**(7), 16195–16203 (2023)
72. M. Hu, W. Wang, W. Cheng, H. Zhang, Initial probability adaptation enhanced cross-entropy-based tone injection scheme for PAPR reduction in OFDM systems. IEEE Trans. Veh. Technol. **70**(7), 6674–6683 (2021). https://doi.org/10.1109/TVT.2021.3078736

Chapter 5
PAPR in MIMO-OFDM System

5.1 Introduction

The exponential increase in demand for high-speed wireless communication is the manifestation of the advancement in wireless broadband technology throughout the years. Multiple-input multiple-output (MIMO) is a technology that offers high quality of service (QoS) with increased spectral efficiency and data rates. The combination of MIMO and OFDM technologies called MIMO-OFDM has become the foundation for next-generation communication systems, ranging from wireless LAN to broadband access [1]. The association of these two innovative technologies could avoid the need for additional bandwidth, which is impractical and costly. Technically, MIMO refers to an antenna configuration that exists in communication systems and can be categorized into four categories with reference to the number of antennas configured at both transmitter and receiver. The first category is the single-input single-output (SISO) system which consists of one transmit and one receive antenna. The second category is the single-input multiple-output (SIMO) system, which consists of one transmit antenna and Q_R receive antenna ($Q_R > 1$). On the other hand, the third category is the multiple-input single-output (MISO) system, which consists of P_T transmit antenna and one receive antenna ($P_T > 1$). Finally, the fourth category is the MIMO system, which consists of P_T transmit antenna and Q_R receive antenna ($P_T, Q_R > 1$).

Apart from the antenna configurations, there are two ways to transmit data across the given channel in the MIMO system due to the nature of the MIMO itself. First, the presence of several antennas in the system implies that numerous paths of propagation exist. The transmission of the same information data across the different propagation paths is called the diversity technique, in which its primary purpose is to improve the reliability of the system. On the other hand, the term spatial multiplexing technique (SM) means transmitting different portions of the information on different propagation paths to improve the system's data rate. Bell Laboratories Layer Space–Time (BLAST) is an example of multiple-antenna transceiver architecture that offers

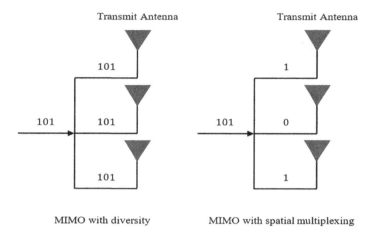

Fig. 5.1 Comparison of diversity and spatial multiplexing technique

spatial multiplexing in the wireless communication system [2]. To get a clear picture of how data is transmitted in MIMO, Fig. 5.1 shows the difference between diversity and spatial multiplexing techniques.

5.2 MIMO-OFDM System

In the MIMO-OFDM system, several signals carrying the same information data simultaneously are transmitted through multiple antennas while simultaneously utilizing a single radio channel, as shown in Fig. 5.2. The utilization of multiple antennas at both transmitter and receiver parts forms an antenna diversity which helps to improve the signal quality and the transmission link. The data is split into several data streams at the transmitter and recombined at the receiver by another MIMO configured with the same antennas.

The MIMO system introduces redundancy into data transmission that classic single antenna configuration SISO could not provide by transmitting the same data on multiple streams. This gives MIMO systems several advantages over typical SISO configurations, such as increased data capacity and improved BER performance.

Despite all the advantages that MIMO-OFDM can offer, it still suffers from a high PAPR issue inherited from the OFDM system. Unconstrained signaling technique such as BLAST applied to the MIMO-OFDM system is intended to increase data rate through simultaneous transmission by using spatial multiplexing over multiple transmission antennas. The BLAST provides a multiplexing gain but lacks diversity gain. This implies that each antenna is independent, which is similar to the SISO scenario [1]. The conclusion that can be drawn is that BLAST in the MIMO-OFDM system has an identical PAPR value to SISO for each antenna in the time domain.

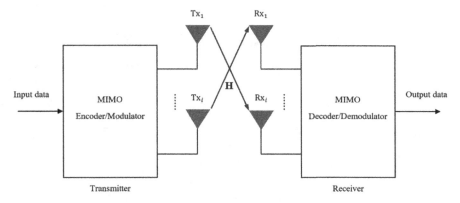

Fig. 5.2 MIMO system (adapted from [3])

The worst-case scenario that could happen is when all the transmit antennas possess a high PAPR value.

5.3 PAPR in MIMO-OFDM System

Suppose a MIMO-OFDM system utilizing spatial multiplexing schemes such in Fig. 5.3, having P_m transmit antenna and Q_n receive antenna. For m th transmit antenna, the OFDM symbol $Y_{m,k} = [Y_{m,0}, Y_{m,1}, \ldots, Y_{m,N-1}]$ where $0 \leq k \leq N - 1$ are considered to be independent [4]. Subsequently, the summation of IFFT modulated subcarriers for m th transmit antenna can be written as [5]

$$y_m(t) = \frac{1}{\sqrt{N}} \sum_{k=0}^{N-1} Y_{m,k} \cdot e^{j2\pi f_k t} 0 < t < NT \quad (5.1)$$

Hence, the PAPR of the transmitted signal for n th transmit antenna is expressed as [5]

$$\text{PAPR}_m = 10\log\left(\frac{max|y_m(t)|^2}{E|y_m(t)|^2}\right) \text{dB} \quad (5.2)$$

As mention above, the worst case of PAPR is considered for all transmitting antennas in a spatial multiplexing scheme. Thus, the PAPR of the spatial multiplexing in the MIMO-OFDM system is given by [5]

$$\text{PAPR}_{MIMO} = \max(\text{PAPR}_m) = \max(\text{PAPR}_1, \text{PAPR}_2, \ldots, \text{PAPR}_{P_m})) \quad (5.3)$$

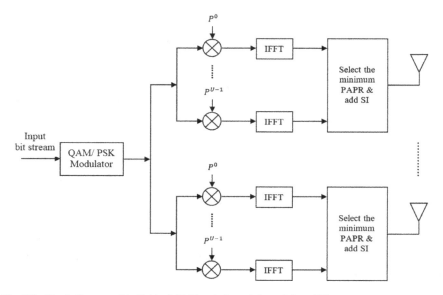

Fig. 5.3 Block diagram of individual SLM technique (adapted from [9])

From Eqs. (5.1) and (5.2), it can be seen that the PAPR of spatial multiplexing MIMO-OFDM system is straightforward from the SISO case. Thus, as long as it bears the same number of subcarriers, N, it inherits the same value of PAPR CCDF as in SISO for every single antenna.

Even though it is possible to easily extend previous PAPR reduction techniques in SISO systems to MIMO-OFDM systems, it would be such a waste as the simple extensions do not take full advantage of the additional level of freedom offered by multiple transmit antenna [4, 6]. Hence, truly understanding the concept and theory of MIMO-OFDM can help adopt the previous technique in SISO to reduce the high PAPR exhibited by the MIMO-OFDM systems [7].

For that reason, two methods, namely individual and concurrent SLM-based MIMO-OFDM system, have been studied in [8]. For individual SLM, the SLM is performed individually on each antenna for two transmit antenna configurations as illustrated in Fig. 5.3. However, by doing so, each OFDM signal from both antennas requires a different source of phase factor, including the use of different side information. In addition, this approach causes the receiver to sustain a decoding complexity that contributes to BER degradation performance.

Unlike individual SLM techniques, the concurrent SLM technique assigned each phase factor to multiple antennas. This implies that concurrent SLM requires the use of the same phase factor to all its transmit antennas. However, this approach is reported to degrade PAPR performance compared with the individual SLM method slightly. Nevertheless, using the same side information for all antennas can reduce the probability of erroneous side information occurring in this technique by taking advantage of the diversity features [10]. As stated by [8], the higher reliability of

5.3 PAPR in MIMO-OFDM System

side information due to the use of the same side information on different antennas provides a spatial diversity in which the decoding process of side information can be made very accurately and reduces the decoding complexity of the receiver. The concurrent SLM method is also being studied in [9, 11].

Now comes the next issue that is which set of the antenna should be selected for transmission. In [8], the authors used the minimum average (mini-average) criterion. The optimal set of signals that gives the least value of PAPR is selected for transmission in the concurrent SLM technique. Anyway, another criterion can be applied in concurrent SLM proposed by [12] called the minimax criterion. Unlike the mini-average criterion, the minimax criterion must first find the maximum PAPR in each set of signals. Then, the minimum set among the maximum of the set that has been identified is selected for transmission. Figure 5.4 illustrates the block diagram of the concurrent SLM technique using a minimax criterion. The simulation result reveals that the minimax criterion is far better in terms of PAPR improvement than the mini-average criterion. Most of the current studies of interleaving and SLM technique in MIMO-OFDM system utilize minimax criterion due to its accurate decoding of side information with lower complexity [13, 14].

Another issue raised in the MIMO-OFDM system is that PAPR calculation occurs at every time slot of block data. For SISO, it is not a major problem as the data sequence is arranged in one block of data for one particular period, similar to spatial multiplexing MIMO-OFDM. However, the case is different in diversity MIMO-OFDM, in which the type of diversity scheme that is used needs to be identified first. For instance, the data symbols are transmitted via multiple antennas at different time slots in STBC and STFBC, exploiting space diversity and time diversity [9]. Besides, such a mapping scheme would require two blocks of data, and thus the period will be divided into two time slots [8, 15]. According to Lee et al. in [8],

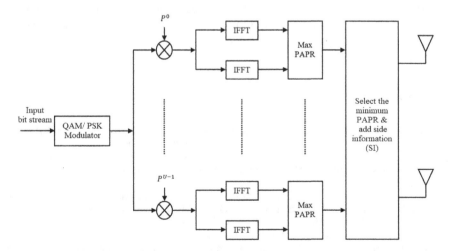

Fig. 5.4 Block diagram of concurrent SLM technique (adapted from [11])

the same PAPR properties shared by s_i and $\pm s_i^*$ ($i = 1,2$) in space–time diversity allow the PAPR calculation to be done only for the first time period [12, 15]. While for SFBC, one-time slot is required to transmit the data symbol through multiple antennas with different frequencies [14, 16, 17].

Since signal behavior is the source of the occurrence of high PAPR, thus, the mapping of the symbol should be emphasized. Alamouti scheme is a commonly used mapping symbol for the STC technique [18]. The mapping of the OFDM symbols for different diversity schemes in the MIMO-OFDM system using the Alamouti scheme will be discussed comprehensively in the next section. Also, the discussion focuses on the diversity combining technique as one of the ways to improve the BER performance.

5.4 MIMO-OFDM System for 5G Networks

With the growing need for high data rates and dependable connection, particularly in the context of 5G networks, the evolution of these approaches will significantly influence the future of wireless communication. Massive MIMO employs a high number of antennas at the base station to simultaneously serve numerous customers, which can result in a significant increase in PAPR owing to both constructive and destructive interference among transmitted signals. High PAPR levels can cause hardware, such as power amplifiers, to become less efficient and more susceptible to breakdown. The PAPR reduction of massive MIMO network communications could significantly enhance system efficiency [19]. Implementing previously developed PAPR reduction strategies in massive MIMO-OFDM systems is critical for improving performance and assuring efficient power amplifier operation. This technology is essential to 5G and beyond 5G networks. It is a promising technology that offers significant advantages in terms of capacity, spectrum efficiency, energy efficiency, and coverage.

References

1. H. Bölcskei, MIMO-OFDM wireless systems: basics, perspectives, and challenges. IEEE Wirel. Commun. **13**(4), 31–37 (2006). https://doi.org/10.1109/MWC.2006.1678163
2. G.J. Foschini, Layered space-time architecture for wireless communication in a fading environment when using multi-element antennas. Bell Labs Tech. J. **1**(2), 41–59 (1996). https://doi.org/10.1002/bltj.2015
3. S. Chirag R, Performance and comparative analysis of SISO, SIMO, MISO, MIMO. Int. J. Wirel. Commun. Simul. **9**(1), 1–14 (2017). [Online]. Available: http://www.ripublication.com
4. Y.H. You, W.G. Jeon, J.H. Paik, Investigation of peak-to-average power ratio in STBC-OFDM. Electron. Lett. **41**(2), 40–41 (2005). https://doi.org/10.1049/el20030654
5. B.R. Karimi, M. Beheshti, M.J. Omidi, PAPR reduction in MIMO-OFDM systems: spatial and temporal processing. Wirel. Pers. Commun. **79**(3), 1925–1940 (2014). https://doi.org/10.1007/s11277-014-1965-y

References

6. H. Zhang, D.L. Goeckel, Peak power reduction in closed-loop MIMO-OFDM systems via mode reservation. IEEE Commun. Lett. **11**(7), 583–585 (2007). https://doi.org/10.1109/LCOMM.2007.070275
7. P. Sunil Kumar, M.G. Sumithra, E. Praveen Kumar, P.G. Scholars, Performance analysis of PAPR reduction in STBC MIMO-OFDM system, in *2013 5th International Conference on Advanced Computing, ICoAC 2013* (2014), pp. 132–136. https://doi.org/10.1109/ICoAC.2013.6921939
8. Y. Lee, Y. You, W. Jeon, J. Paik, H. Song, Peak-to-average power ratio in MIMO-OFDM systems using selective mapping. IEEE Commun. Lett. **7**(12), 575–577 (2003)
9. E.S. Hassan, S.E. El-Khamy, M.I. Dessouky, S.A. El-Dolil, F.E. Abd El-Samie, Peak-to-average power ratio reduction in space-time block coded multi-input multi-output orthogonal frequency division multiplexing systems using a small overhead selective mapping scheme. IET Commun. **3**(10), 1667–1674 (2009). https://doi.org/10.1049/iet-com.2008.0565
10. B.M. Lee, R.J.P. de Figueiredo, MIMO-OFDM PAPR reduction by selected mapping using side information power allocation. Digital Signal Process. Rev. J. **20**(2), 462–471 (2010). https://doi.org/10.1016/j.dsp.2009.06.025
11. T. Jiang, C. Ni, L. Guan, A novel phase offset SLM scheme for PAPR reduction in Alamouti MIMO-OFDM systems without side information. IEEE Signal Process. Lett. **20**(4), 383–386 (2013). https://doi.org/10.1109/LSP.2013.2245119
12. M. Tan, Z. Latinovic, Y. Bar-Ness, STBC MIMO-OFDM peak-to-average power ratio reduction by cross-antenna rotation and inversion. IEEE Commun. Lett. **9**(7), 592–594 (2005). https://doi.org/10.1109/lcomm.2005.1461674
13. S. Umeda, S. Suyama, H. Suzuki, K. Fukawa, PAPR reduction method for block diagonalization in multiuser MIMO-OFDM systems, in *IEEE Vehicular Technology Conference* (2010), pp. 1–5. https://doi.org/10.1109/VETECS.2010.5493834
14. N. Taspinar, M. Yildirim, A novel parallel artificial bee colony algorithm and its PAPR reduction performance using SLM scheme in OFDM and MIMO-OFDM systems. IEEE Commun. Lett. **19**(10), 1830–1833 (2015). https://doi.org/10.1109/LCOMM.2015.2465967
15. M. Sghaier, F. Abdelkefi, M. Siala, PAPR reduction scheme with efficient embedded signaling in MIMO-OFDM systems. EURASIP J. Wirel. Commun. Netw. **2015**(1), 1–16 (2015). https://doi.org/10.1186/s13638-015-0443-x
16. H.B. Jeon, J.S. No, D.J. Shin, A low-complexity SLM scheme using additive mapping sequences for PAPR reduction of OFDM signals. IEEE Trans. Broadcast. **57**(4), 866–875 (2011). https://doi.org/10.1109/TBC.2011.2151570
17. Y.S. Su, Computationally efficient PAPR reduction of SFBC-OFDM signals by polyphase interleaving and inversion, in *International Symposium on Wireless Personal Multimedia Communications, WPMC*, vol. 2015-Janua, no. 1 (2015), pp. 729–733. https://doi.org/10.1109/WPMC.2014.7014911
18. S.M. Alamouti, A simple transmit diversity technique for wireless communications. IEEE J. Sel. Areas Commun. **16**(8), 1451–1458 (1998)
19. C.A. Schmidt, J.F. Schmidt, J.L. Figueroa, M. Crussiere, Achievable energy efficiency in massive MIMO: impact of DAC resolution and PAPR reduction for practical network topologies at mm-waves. IEEE Commun. Lett. **26**(11), 2784–2788 (2022). https://doi.org/10.1109/LCOMM.2022.3198016

Chapter 6
Diversity Techniques

6.1 Introduction

Diversity scheme is a technique that efficiently and conveniently combats the destructive effect of fading channels in the wireless communication system. This scheme is capable of exploiting the random aspect of radio propagation [1]. The fundamental principle of diversity is the repetition or redundancy of information. Several transmitted signal paths are likely to have a low probability of experiencing deep fades simultaneously. Selecting the antenna with the strongest signal will result in a much better signal than the one the antenna can achieve. Thus, having more than one path will significantly improve the receiver's instantaneous and average SNR [2, 3]. A variety of diversity techniques are widely used in wireless communication systems, such as space diversity, time diversity, and frequency diversity [3, 4]. This research will focus on the space–time code (STC) scheme, including space–time, space-frequency, and space-frequency.

The introduction of a diversity scheme in the PAPR reduction technique becomes a solution to improving BER performance. The diversity gains through the implementation of space–time block code (STBC), space-frequency block code (SFBC), and space–time-frequency block code (STFBC) help to enhance the performance and reliability of the system. Although the OFDM has benefited from side information (SI) in the previously proposed technique to improve its BER degradation in SISO, the implementation of the diversity scheme will allow the BER performance in the system to be further improved [5].

There are two main subcarrier mapping modes in the STC scheme: space–time arrangement and space-frequency arrangement. STBC exploits the spatial (antennas) and time in which the encoding process is done in antenna/time domains [4]. In contrast, SFBC exploits the spatial (antennas) and frequency, enabling the encoding process in antenna/frequency domains. The Alamouti scheme is the common data mapping used in STBC and SFBC encoding processes. The arrangement of symbols into two groups becomes the basis of the Alamouti scheme and is usually represented

Fig. 6.1 STBC encoding using Alamouti scheme

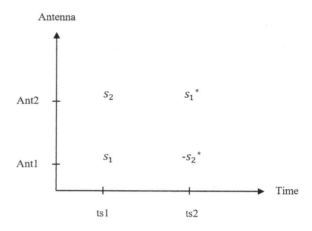

by a matrix [6]:

$$\begin{bmatrix} s_1 & s_2 \\ -s_2^* & s_1^* \end{bmatrix} \quad (6.1)$$

The time slot information/frequency and the antenna from which the signal is sent are represented by the rows and column respectively. In STBC encoding, at time t, antenna 1 transmits signal s_1 and antenna 2 transmits signal s_2, and, similarly, at time $t + T$, antenna 1 transmits signal $-s_2^*$ and antenna 2 transmits signal s_1^*. The input of STBC encoder and decoder is in the form of a block of transmission symbols. The distribution of the block data is illustrated in Fig. 6.1. The signal which originally owns one transmission path now has four new transmission paths, thanks to the exploitation of two antennas and two time slots.

Subsequently, the signal will undergo multiplication of channel distribution, H [6]:

$$H = \begin{bmatrix} h_1 & h_2 \end{bmatrix} \quad (6.2)$$

Hence, the received signal with noise can be written as [6]

$$r = H * s + n \quad (6.3)$$

For one received antenna in diversity scheme, the received signal after the FFT modulation can be expressed as [6]

$$r_1 = r(t) = h_1 s_1 + h_2 s_2 + n_1 \quad r_2 = r(t + T) = -h_1 s_2^* + h_2 s_1^* + n_2 \quad (6.4)$$

where the received signal at time slot t is represented by r_1 while the received signal at the next time slot $t + T$ is represented by r_2. Thus, by using the maximum ratio

6.1 Introduction

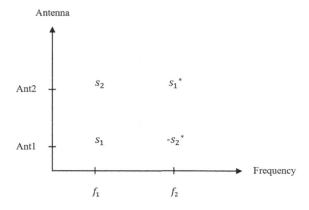

Fig. 6.2 SFBC encoding using Alamouti scheme

combiner (MRC) technique, the retrieved data can be expressed as [7]

$$\tilde{s}_1 = h_1^* r_1 + h_2 r_2^* \\ \tilde{s}_2 = h_2^* r_1 - h_1 r_2^* \quad (6.5)$$

where $\tilde{s}_1$ and $\tilde{s}_2$ denote the estimation signal obtained after the decoding process, while for the SFBC diversity scheme, SF coding aims to provide high diversity to the wireless communication system, which improves efficiency by leveraging the frequency and antenna. Figure 6.2 represents the distribution of the SFBC encoding block data. There is only one time slot in SFBC with different antennas and frequencies.

The symbol codeword of SFBC MIMO-OFDM can be expressed as [8, 9]

$$C = \begin{bmatrix} X_{0,1} & \cdots & X_{0,N_t} \\ X_{0,1} & \cdots & X_{0,N_t} \\ \vdots & \cdots & \vdots \\ X_{N-1,1} & \cdots & X_{N-1,N_t} \\ X_{N-1,1} & \cdots & X_{N-1,N_t} \end{bmatrix} \quad (6.6)$$

where N and N_t represent the number of subcarriers and the number of transmit antennas respectively. Based on Eq. (6.6), it can be shown that the formation of SF code is through repetition of each row r times.

The effectiveness of previous space–time and space-frequency coding system has led to space–time-frequency (STF) coding development. In STFBC schemes, a frequency dimension is combined with space and time dimensions as shown in Fig. 6.3.

Similar to STBC and SFBC, this approach can enhance system reliability and performance and achieve maximum diversity gain [9, 10]. STFBC can generate

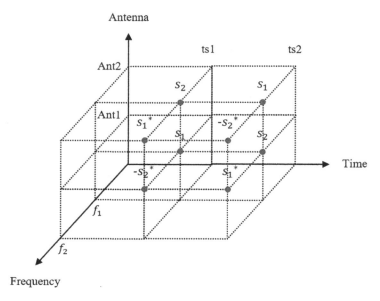

Fig. 6.3 STFBC encoding using Alamouti scheme

eight independent signal paths equivalent to twice the signal path available in STBC and SFBC.

6.2 Diversity Reception Combining

In receiver diversity, where multiple receive antennas receive multiple signals from multiple independent fading paths, one method must combine those signals. This is where the diversity combining technique comes in handy. As its name indicates, the diversity combining technique is a method to combine the diversity received signal at the receiver into a single improved signal [11]. Diversity has long been recognized as an effective technique for enhancing the reliability of wireless communication systems by reducing the detrimental effects of channel fading [6]. The key idea of diversity is to maximize the received signal-to-noise ratio (SNR) through extracted information from the received signal components transmitted over multiple fading channels [12]. Technically, the signals received by the receiver is the same. Still, the only difference is the level of fading experienced by every one of them, with the possibility that all these duplicates being in a deep fade is slightly low. Thus, the existence of diversity combining technique is to serve the purpose of counteracting low scale fading. Therefore, in this research, the slow, flat Rayleigh fading is used as the channel model for the signal fluctuation as it is the simplest and most facile model.

6.2 Diversity Reception Combining

The physical model expects the fading channel to be independent of one factor to another. Thus, it can be treated as an independent sample of the random fading process. Then, an independent copy of the transmitted signal is assigned to each sequence factor. It is safe to say that not all factors experienced a deep fade, implying at least one factor not severely affected by the deep fade. It possesses a reasonable power to be used by the receiver for further process [6]. Thus, the sample of a fading channel can be expressed as [6]

$$H = \begin{bmatrix} h_{1,1} & \cdots & h_{1,N_r} \\ \vdots & \ddots & \vdots \\ h_{N_t,1} & \cdots & h_{N_t,N_r} \end{bmatrix} \quad (6.7)$$

For the 2×2 MIMO system, the signal receives by antenna one as given by [6]

$$\begin{bmatrix} r_1 \\ r_2 \end{bmatrix} = \begin{bmatrix} s_1 & s_2 \\ -s_2^* & s_1^* \end{bmatrix} \cdot \begin{bmatrix} h_{1,1} \\ h_{2,1} \end{bmatrix} + \begin{bmatrix} n_1 \\ n_2 \end{bmatrix} \quad (6.8)$$

and the signal receives by antenna two as given by [6]

$$\begin{bmatrix} r_3 \\ r_4 \end{bmatrix} = \begin{bmatrix} s_1 & s_2 \\ -s_2^* & s_1^* \end{bmatrix} \cdot \begin{bmatrix} h_{1,2} \\ h_{2,2} \end{bmatrix} + \begin{bmatrix} n_3 \\ n_4 \end{bmatrix} \quad (6.9)$$

Three types of diversity combining techniques were used practically in the wireless communication system, which is selection combining (SC), maximal ratio combining (MRC), and equal gain combining (EGC). These techniques share the same purpose, which is to seek a set of weights $\hat{w} = [w_1, w_2, \ldots, w_z]$ as depicted in Fig. 6.4, optimizing a specific objective function. The weights are selected to minimize the effect of fading on the received multiple signal components for a single user. The selection process varies depending on what scheme is used. In this research, it is assumed that the receiver has the required knowledge of channel fading, H.

As a simple diversity technique, the selection process in the selection combining is very straightforward in which the branch with the highest SNR is selected. This is essentially the classical form of diversity communication [6, 13]. Therefore, the selection criterion of weight can be written as [6]

$$w_m = \begin{cases} 1, & \text{if } \epsilon_m = max_n\{\epsilon_n\} \\ 0, & \text{if otherwise} \end{cases} \quad (6.10)$$

where ϵ_m is the largest SNR in the selection combining scheme. This scheme would require only one signal power calculation where the other factors were explicitly overlooked. The simulation results demonstrate that this approach can barely offer an optimal solution [6].

In MRC [14], the received signals are weighted accordingly so that the SNR at the output of the combiner is the sum of the average SNR of each branch.

Fig. 6.4 Diversity reception combining (adapted from [12])

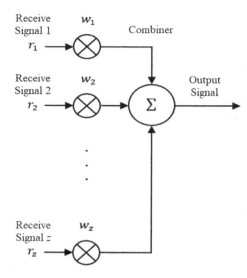

Mathematically, the weight of each received signals is given as [15]

$$r(t) = H(t) * s(t) + n \quad (6.11)$$

$$H = [h_1, h_2, \ldots, h_z]^T \quad (6.12)$$

$$n = [n_1, n_2, \ldots, n_z]^T \quad (6.13)$$

$$r(t) = w^H \cdot r = w^H \cdot H(t) * s(t) + w^H \cdot n \quad (6.14)$$

The instantaneous output SNR is given by [15]

$$\epsilon = \frac{|w^H \cdot H|^2}{E\{|w^H \cdot n|^2\}} \quad (6.15)$$

where the noise power $E\{|w^H \cdot n|^2\}$ can be derived as [15]

$$\begin{aligned} E\{|w^H \cdot n|^2\} &= E\{|w^H n n^H w|\} \\ &= w^H E\{n n^H\} w \\ &= \sigma^2 w^H I_N w \\ &= \sigma^2 w^H w \\ &= \sigma^2 \|w^2\| \end{aligned} \quad (6.16)$$

6.2 Diversity Reception Combining

where I_N represents an $N \times N$ identity matrix. By assuming that $\|w\| = 1$, therefore, the SNR of the received signal is given as [15]

$$\epsilon = \frac{|w^H \cdot H|^2}{\sigma^2} \qquad (6.17)$$

Therefore, the maximum SNR that can be achieved with conditions w must be linearly proportional to H [15]:

$$w = H$$

$$\therefore \epsilon = \frac{|H^H \cdot H|^2}{\sigma^2 H^H \cdot H} = \frac{H^H \cdot H}{\sigma^2} = \sum_{n=0}^{N-1} \frac{|h_n|^2}{\sigma^2} = \sum_{n=0}^{N-1} \epsilon_n \qquad (6.18)$$

Meanwhile in EGC, the received signals are weighted equally and then added, thus making this approach easier to implement that MRC. EGC forms a unit gain at each branch to maximize its average SNR in the system. Therefore, the weight, w, can be given by [12]

$$w_z = e^{j \angle h_z} \Rightarrow w_z * h_z = |h_z| \Rightarrow w \cdot H = \sum_{n=0}^{N-1} |h_z| \qquad (6.19)$$

The noise power and instantaneous SNR can be written as [12]

$$P_n = w^H w \sigma^2 = N\sigma^2 \qquad (6.20)$$

$$\epsilon = \frac{\gamma^2}{N\sigma^2}, \gamma = \sum_{n=0}^{N-1} |h_z| \qquad (6.21)$$

From the three combiners discussed above, each one of them has its advantages. However, when it comes to the best performance in Rayleigh fading channel, MRC is the one [16] while the worst one is the SC technique [16–18]. The high diversity gain and high signal-to-noise ratio owned by the MRC have made it superior to other diversity techniques [16].

References

1. P. Sunil Kumar, M.G. Sumithra, E. Praveen Kumar, P.G. Scholars, Performance analysis of PAPR reduction in STBC MIMO-OFDM system, in *2013 5th International Conference on Advanced Computing, ICoAC 2013* (2014), pp. 132–136. https://doi.org/10.1109/ICoAC.2013.6921939
2. Y. Lee, Y. You, W. Jeon, J. Paik, H. Song, Peak-to-average power ratio in MIMO-OFDM systems using selective mapping. IEEE Commun. Lett. **7**(12), 575–577 (2003)
3. E.S. Hassan, S.E. El-Khamy, M.I. Dessouky, S.A. El-Dolil, F.E. Abd El-Samie, Peak-to-average power ratio reduction in space-time block coded multi-input multi-output orthogonal frequency division multiplexing systems using a small overhead selective mapping scheme. IET Commun. **3**(10), 1667–1674 (2009). https://doi.org/10.1049/iet-com.2008.0565
4. B.M. Lee, R.J.P. de Figueiredo, MIMO-OFDM PAPR reduction by selected mapping using side information power allocation. Digital Signal Process. Rev. J. **20**(2), 462–471 (2010). https://doi.org/10.1016/j.dsp.2009.06.025
5. T. Jiang, C. Ni, L. Guan, A novel phase offset SLM scheme for PAPR reduction in Alamouti MIMO-OFDM systems without side information. IEEE Signal Process. Lett. **20**(4), 383–386 (2013). https://doi.org/10.1109/LSP.2013.2245119
6. S.M. Alamouti, A simple transmit diversity technique for wireless communications. IEEE J. Sel. Areas Commun. **16**(8), 1451–1458 (1998)
7. M. Tan, Z. Latinovic, Y. Bar-Ness, STBC MIMO-OFDM peak-to-average power ratio reduction by cross-antenna rotation and inversion. IEEE Commun. Lett. **9**(7), 592–594 (2005). https://doi.org/10.1109/lcomm.2005.1461674
8. S. Umeda, S. Suyama, H. Suzuki, K. Fukawa, PAPR reduction method for block diagonalization in multiuser MIMO-OFDM systems, in *IEEE Vehicular Technology Conference* (2010), pp. 1–5. https://doi.org/10.1109/VETECS.2010.5493834
9. N. Taspinar, M. Yildirim, A novel parallel artificial bee colony algorithm and its PAPR reduction performance using SLM scheme in OFDM and MIMO-OFDM systems. IEEE Commun. Lett. **19**(10), 1830–1833 (2015). https://doi.org/10.1109/LCOMM.2015.2465967
10. M. Sghaier, F. Abdelkefi, M. Siala, PAPR reduction scheme with efficient embedded signaling in MIMO-OFDM systems. EURASIP J. Wirel. Commun. Netw. **2015**(1), 1–16 (2015). https://doi.org/10.1186/s13638-015-0443-x
11. H.B. Jeon, J.S. No, D.J. Shin, A low-complexity SLM scheme using additive mapping sequences for PAPR reduction of OFDM signals. IEEE Trans. Broadcast. **57**(4), 866–875 (2011). https://doi.org/10.1109/TBC.2011.2151570
12. W. Su, Z. Safar, K.J.R. Liu, Towards maximum achievable diversity in space, time, and frequency: performance analysis and code design. IEEE Trans. Wirel. Commun. **4**(4), 1847–1857 (2005). https://doi.org/10.1109/TWC.2005.850323
13. W. Zhang, X. Gen, K. Ben Letaief, Space-time/frequency coding for MIMO-OFDM in next generation broadband wireless systems. IEEE Wirel. Commun. **14**(3), 32–43 (2007)
14. A. Lozano, N. Jindal, Transmit diversity vs. spatial multiplexing in modern MIMO systems. IEEE Trans. Wirel. Commun. **9**(1), 186–197 (2010). https://doi.org/10.1109/TWC.2010.01.081381
15. V. Tarokh, N. Seshadri, A.R. Calderbank, Space-time codes for high data rate wireless communication: performance criterion and code construction. IEEE Trans. Inf. Theory **44**(2), 744–765 (1998). https://doi.org/10.1109/18.661517

References

16. K. Pachori, A. Mishra, An efficient combinational approach for PAPR reduction in MIMO–OFDM system. Wirel. Netw. **22**(2), 417–425 (2016). https://doi.org/10.1007/s11276-015-0974-4
17. V. Ahirwar, B.S. Rajan, Low PAPR full-diversity space-frequency codes for MIMO-OFDM systems. GLOBECOM IEEE Global Telecommun. Conf. **3**, 1476–1480 (2005). https://doi.org/10.1109/GLOCOM.2005.1577896
18. B. Ripan, K. Roy, T.K. Roy, BER analysis of MIMO-OFDM system using Alamouti STBC and MRC diversity scheme over Rayleigh multipath channel. Global J. Res. Eng. Electr. Electron. Eng. **13**(13), 14–24 (2013)

Index

A
Access
  high-speed packet data access (HSPA), 2

B
Bandwidth, 1, 7, 9, 12, 40, 45
Bell Laboratories Layer Space-Time (BLAST), 45, 46
Bit Error Rate (BER), 3, 8, 14, 24, 25, 27–32, 34–36, 38–40
Block code
  space frequency block code (SFBC), 50, 53, 55
  space time block code (STBC), 49, 53–55
  space time frequency block code (STFBC), 49, 53, 55

C
Carrier
  multicarrier signal, 3
  receiver carrier, 11
Channel
  subchannels, 1
  uplink & downlink, 13
Computational complexity, 3, 27, 30–36
Conversion
  serial-to-parallel, 9, 37
Converter
  analog-to-digital, 11
  digital-to-analog, 10, 14
Cyclic prefix, 11

D
Demodulation, 11, 20–22
Distortion
  in-band, 13, 14, 25, 27, 28
  multi-path, 10
Domain
  frequency, 11, 21, 53
  time, 11, 19, 21, 32, 34, 37, 46, 53

E
Efficiency
  energy, 2, 7, 8, 13
  power, 2, 8, 13–15
  spectral, 1, 2, 7–9, 24, 28
  spectrum, 2, 7, 50
Euclidean, 34

F
Fading
  frequency selective, 2
  Rayleigh, 23, 25, 56

G
GreenOFDM, 34, 35
Guard interval, 10, 11

H
High
  high data transmission, 9
  high-power amplifier (HPA), 2, 8, 11, 13–15, 25, 27

I

Interference
adjacent channel, 13
in-band, 27
out-band, 27
Inverse Fast Fourier Transform (IFFT), 9,
10, 13, 19–21, 31–33, 35–38, 40, 47

K

Known Symbol Padding (KSP), 11

M

Modulation
binary phase shift keying (BPSK), 10,
12, 35
digital, 9, 12, 24, 31
phase shift keying (PSK), 19, 24
quadrature amplitude modulation
(QAM), 10, 12, 19, 24, 25, 35, 37
quadrature phase shift keying (QPSK),
10, 12
Multiple-in Multiple-out (MIMO)
MIMO-OFDM, 2, 3, 29, 38, 45–50, 55
Multiplexing
spatial, 3, 45, 47

O

Orthogonal
orthogonal frequency division
multiplexing (OFDM), 7, 8

P

Path loss, 2
Peak-to-Average Power Ratio (PAPR), 2, 3,
7, 8, 13–15, 22, 23, 25, 27–32,
34–36, 38, 46, 47
Permutation, 3, 30, 31
Propagation
multipath, 1

Q

Quasi
quasi cyclic LDPC (QC-LDPC), 30

S

Scheme
alamouti, 50, 53
companding, 27
Sequence
symbol, 13, 34, 36
Side information, 3, 25, 27, 30, 31, 33–36,
38, 48, 49
Single Input Single Output (SISO), 3, 45,
46

T

Technique
block coding, 28, 29
clipping, 27
companding, 27, 39
discrete cosine transforms (DCT), 39,
40
discrete hartley transform (DHT), 39, 40
interleaving, 3, 31, 32, 36
partial transmit sequence (PTS), 40
scrambling, 30
selective codeword shift (SCS), 36–38
selective mapping technique (SLM), 3,
30–37, 40, 48
Walsh-Hadamard transform (WHT), 39,
40
Turbo codes, 28–30

U

Ultra-Wideband (UWB), 11

V

Vertical Bell Laboratory Layered
Space-Time (V-BLAST), 3
Very high-speed digital subscriber line, 11

W

Wideband Code Division Multiple Access
(WCDMA)
access network, 2
channel, 1, 7
WiMAX, 2, 11
WLAN, 11

Z

Zero padding, 11

Printed in the United States
by Baker & Taylor Publisher Services